797,885 Books

are available to read at

Forgotten Books

www.ForgottenBooks.com

Forgotten Books' App
Available for mobile, tablet & eReader

Download on the
App Store

ANDROID APP ON
Google play

ISBN 978-1-331-30297-1
PIBN 10171498

This book is a reproduction of an important historical work. Forgotten Books uses
state-of-the-art technology to digitally reconstruct the work, preserving the original format
whilst repairing imperfections present in the aged copy. In rare cases, an imperfection in
the original, such as a blemish or missing page, may be replicated in our edition. We do,
however, repair the vast majority of imperfections successfully; any imperfections that
remain are intentionally left to preserve the state of such historical works.

Forgotten Books is a registered trademark of FB &c Ltd.
Copyright © 2015 FB &c Ltd.
FB &c Ltd, Dalton House, 60 Windsor Avenue, London, SW19 2RR.
Company number 08720141. Registered in England and Wales.

For support please visit www.forgottenbooks.com

1 MONTH OF
FREE
READING

at
www.ForgottenBooks.com

By purchasing this book you are eligible for one month membership to ForgottenBooks.com, giving you unlimited access to our entire collection of over 700,000 titles via our web site and mobile apps.

To claim your free month visit: www.forgottenbooks.com/free171498

* Offer is valid for 45 days from date of purchase. Terms and conditions apply.

English
Français
Deutsche
Italiano
Español
Português

www.forgottenbooks.com

Mythology Photography **Fiction**
Fishing Christianity **Art** Cooking
Essays Buddhism Freemasonry
Medicine **Biology** Music **Ancient
Egypt** Evolution Carpentry Physics
Dance Geology **Mathematics** Fitness
Shakespeare **Folklore** Yoga Marketing
Confidence Immortality Biographies
Poetry **Psychology** Witchcraft
Electronics Chemistry History **Law**
Accounting **Philosophy** Anthropology
Alchemy Drama Quantum Mechanics
Atheism Sexual Health **Ancient History**
Entrepreneurship Languages Sport
Paleontology Needlework Islam
Metaphysics Investment Archaeology
Parenting Statistics Criminology
Motivational

HANDBOOK OF NATURE STUDY

FOR TEACHERS AND PUPILS IN ELEMENTARY SCHOOLS

BY

D. LANGE

INSTRUCTOR IN NATURE STUDY IN THE PUBLIC SCHOOLS
OF ST. PAUL, MINNESOTA

REESE LIBRARY
OF THE
UNIVERSITY
OF CALIFORNIA

New York
THE MACMILLAN COMPANY
LONDON: MACMILLAN & CO., Ltd.
1898

All rights reserved

COPYRIGHT, 1898,

By THE ·MACMILLAN COMPANY.

74999

Norwood Press

J. S. Cushing & Co. — Berwick & Smith
Norwood Mass. U.S.A.

REESE LIBRARY
OF THE
UNIVERSITY
OF CALIFORNIA

PREFACE

THE study of nature with a view to understand the relations of plant and animal life to the welfare and happiness of man, needs no justification in this age of scientific agriculture and applied sciences. All our most progressive teachers agree that Elementary Science or Nature Study should have a place on the programme of every graded and ungraded school in the land.

Purpose of this book. — The writer has attempted to point out some of the material which may be made the basis of profitable lessons in Nature Study, and he has endeavored to show how this material may be made available and what the pupils may be taught about it.

Plan of the book. — A glance at the contents of the different chapters will show that the writer has tried to learn directly from nature, and he would strongly urge teachers and pupils to attend nature's own school in fields and forests, and at lakes and streams. Those happy children who can spend their vacations outdoors do not confine their attention to one class of objects and phenomena. While on a ramble through the woods they naturally observe the birds, the insects, the trees, shrubs, and flowers of the season; and a similar statement is true of lakes and rivers, of swamp, marsh, and meadow. I have, therefore, arranged

the subject-matter according to seasons and life communi-
ties, under such chapters as " Life in and near the Water
in Summer," "The Prairie in Late Summer," "The Woods
in their Autumn Foliage," etc.   An effort has been made
throughout the book to show that all nature is one.   Scientific classification has not been excluded, but has been kept
in the background, because the child must study the con-
crete before he can make abstractions.   Many of our high
school and college students take but a limited interest in
Botany and Zoölogy, because they possess almost nothing
of what may be called "common knowledge about the life
around us."   How can we expect that these young pupils
should suddenly be interested in the microscopic structure
of plants and animals, which they have always passed by
with stupid indifference; or for what possible reason should
scientific systematization attract them, when they know
nothing about the life history of the trees in their yards and
of the birds that nest on them?   This little volume is, however, specially intended to point out work for those children
whose school years close with the common school course.
Their education must remain elementary, but it should,
nevertheless, be broad and thorough.   The writer does not
believe that the physical sciences should be excluded from
our common schools; they should receive special attention
during the last or the last two school years, but it has not
been deemed advisable to include much of them in this
volume.

Suggestions to teachers and parents. — In order to do successfully some of the work suggested and outlined, it is not

necessary to have taken a course in Botany, Zoölogy, or Geology. You can do the work if you have an earnest desire to do it. The attention of the children should be directed to the subject of the lesson some time before the lesson is given. For this purpose teachers should consult the footnotes and the paragraph on the subject under consideration, and plan their work at least a few weeks in advance. Procure as much of the suggested material as possible, and procure other material if it seems helpful for the purpose of the lesson.

Whenever practicable, each pupil should have all or some of the material on his desk; when that is not practicable, the material may be shown to the pupils, or it may be passed around. The facts which you desire to teach should be derived from previous observations and from the material before the class. You may tell the pupils what you cannot direct them to find out for themselves, but it is almost always a bad plan to read to the pupils during Nature Study lessons. Before you give a lesson on an animal or a plant or a phenomenon, you should know just what you want to teach about it; therefore you should always have a carefully prepared outline in your mind.

At the close of the lesson, let one or more pupils sum up in well-constructed sentences the results of the lesson. No broken sentences should be allowed in this review. At the close of a chapter, give a general review as outlined and suggested in the book. A live teacher will find in Nature Study work many good subjects for compositions and much valuable material for drawing.

It is a good plan to have the children collect most of the material, but the teacher should see to it that they do not violate the property rights of others, unnecessarily destroy plants, or treat animals with cruelty. For various reasons, the writer cannot advocate animal dissection in common schools. The only instruments required for plant dissection are a pin and a sharp knife. A small magnifying lens would be valuable to the teacher. The pupils must not use a book for this work during the lesson; but some good book on the subject could profitably be placed in the hands of the older pupils for home reading, or for reviewing topics studied in the class-room. That the teacher should not give Nature Study lessons with an open book before her is self-evident.

It was not and could not be the writer's intention to lay down a programme of Nature Study for any particular school or schools. The term's programme in this work depends so much on the seasons and on local conditions, that the making of it should be left to the teacher, principal, or superintendent. Nor is it the author's idea that any teacher should rigidly adhere to the material and its arrangement in any particular book. If, for instance, the mocking bird is more common near your school than the robin, then study the former instead of the latter. For the methods of collecting a few plants, insects, and other material, teachers are referred to the appendix.

It is believed that the work laid out in this book will furnish enough material for three or four school years of nine months each, if two weekly lessons of about thirty

minutes each are given to it. In ungraded schools the
teacher should form one class of all pupils, or nearly all
pupils, for this subject, but she should not excuse the older
pupils from it. As the opportunities for observations are
best in the country, this work is not at all specially difficult
in ungraded country schools. The writer believes that a
teacher of average ability can make this work profitable
for pupils between the ages of eight and eighteen. The
teacher who has never tried work of this kind will be
surprised how readily pupils take hold of it, provided
that they be taught directly from nature. Nor does the
teacher need to feel disgraced if she cannot answer every
question asked by the children. There are many questions connected with Nature Study which no mortal has
ever answered.

The territory for which this book is especially designed
extends from the Atlantic coast to the Rocky Mountains,
and from the Canadian Provinces to the latitude of southern
Virginia and Kentucky.

Illustrations. — Their main purpose is to assist teachers
in the identification of the plants and animals to be studied.
Nos. 3 and 4 are originals, drawn by Miss Josephine E.
Tilden, of Minneapolis; all the other originals were drawn
by Miss Henrietta G. Fox, of Minneapolis. For those which
are not originals, due credit is given elsewhere.

Nomenclature. — On vertebrates I have followed the fifth
edition of Jordan's "Manual of the Vertebrate Animals
of the Northern United States"; on insects, Comstock's
"Manual for the Study of Insects"; and on plants, the

sixth edition of Gray's "Manual of the Botany of the Northern United States."

Besides consulting well-known writers on Natural History subjects, I have derived much help from the publications of the United States Department of Agriculture, from the publications of the Agricultural Experiment stations of several states, and from the valuable German manuals of Kiessling and Pfalz.

To my fellow-teachers, J. E. Kenny, O. T. Denny, and H. W. Slack, of this city I am indebted for reading the manuscript. Messrs. Kenny and Denny have also rendered me valuable assistance in reading the proof.

With the wish that this little book may do its share of work in the spread of education and in making our boys and girls better and happier, it is submitted to teachers and parents.

D. LANGE.

ST. PAUL, MINNESOTA,
April, 1898.

CONTENTS

PART FIRST

PART SECOND

LIST OF ILLUSTRATIONS

SOURCES OF ILLUSTRATIONS

NUMBERS 9, 10, 11, 12, 19, 26, 27, 28, 29, 30, 31, 32 33, 34, 35, 36, 37, 44, 45, 46, 47, 48, 49, 50, 59, and 60 are originals, drawn from nature by Miss Henrietta G. Fox, of Minneapolis.

Numbers 3 and 4 are originals drawn from nature by Miss Josephine E. Tilden, of Minneapolis.

Numbers 7, 8, 38, 39, 52, and 53, from F. E. Beal's "Some Common Birds in Their Relation to Agriculture." United States Department of Agriculture.

Numbers 54 and 55, from A. K. Fisher's "Hawks and Owls from the Standpoint of the Farmer." United States Department of Agriculture.

Numbers 13, 14, and 15, from Hitchcock and Norton's "Weeds of Kansas, III." In Bulletin No. 57, Kansas State Agricultural College.

Numbers 2, 5, 6, 20, 21, 22, 23, 24, 25, 42, 51, 56, 57, and 58 have been adapted from various sources.

Numbers 16, 17, 18, and 40, from E. S. Goff's "Noxious Weeds." In Bulletin No. 39, Wisconsin Agricultural Experiment Station.

No. 43, from Riley's "Report on the Rocky Mountain Locust."

HANDBOOK OF NATURE STUDY

PART FIRST

I

ABOUT HOME. MARCH TO JUNE

INTRODUCTORY OBSERVATION

§ 1. The warm spring sun has melted the snow, and tiny blades of young grass begin to appear among the dead stalks of last year.   In sheltered, sunny places we notice spiders and insects crawling about, evidently taking their first spring airing.   If you have a flower garden, your tulips will soon be in bloom and your shrubs will have leaves and flowers.   Wild as well as cultivated fruit trees will soon be covered with white or rose-colored flowers, filling the air with fragrance and attracting thousands of buzzing bees.   If we observe carefully, we shall find beetles and caterpillars feeding on the tender foliage, as soon as the buds have opened.   But several kinds of birds have returned to us from the South, and they will probably prevent most of the insects from becoming too numerous.

§ 2. **The Common Tulip**. *Tulipa Gesneriana.*

MATERIAL: Tulips with bulbs ; several sprouting onions.

This plant grows from six to eight inches high.   Each stem carries only one flower and rises from an onion-like

---

Observations. — Catkins of poplars, flowers of willows, eggs of tent-caterpillars.

B

bulb. We shall study this plant somewhat minutely, because it shows nearly all the parts of a plant very clearly.

The beautiful flower consists of six large, white, reddish or yellowish leaves. Before the opening of the flower these leaves are folded around the delicate parts inside, and protect them from frost and cold rains.

Within the bright-colored flower leaves we notice, arranged in a whirl or circle, six tiny bags filled with a yellow dust, which adheres to our fingers, if we touch one of the broken dust bags. The little bags are attached to delicate thread-like bodies, which we shall call dust threads, because they carry the flower dust in the little bags.

In the centre of the flower we find a body, which resembles in shape the pestle of a very small mortar, and it is therefore called the pistil. The head of the pistil looks as if it had been injured and were just healing. This scar-like surface is called the stigma. Now, let us cut open the lower part of the pistil, and we shall find the small, white ovules, or rudimentary seeds. The lower part of the pistil is called the ovary because it contains the ovules.

The stem bears but few leaves, which always arise very close to the ground. They are soft and juicy and surround the stem with their bases. What is their shape? Are they rough, hairy, or smooth?

If we carefully separate the leaves of a growing or sprouting onion, we find that their bases are widened and that each outer leaf completely encircles the next inner one. The fleshy part is enclosed in several dry skins, which keep the moisture in the bulb and also prevent its decay. You may now compare the structure of the tulip bulb with the structure of an onion, and then describe the former.

Observations. — Reappearance of frogs, frog eggs in ponds and ditches, the development of the tadpoles.

§ 3. A Few Garden Vegetables.

MATERIAL: Seeds of the bean, pea, cucumber, melon, squash, pumpkin; pickled cucumbers, tomatoes, and cauliflower; lettuce, cabbage, kohlrabi, asparagus, white celery, turnip, radish, horseradish, carrots, rutabagas. See: Some Additions to Our Vegetable Dietary, by F. V. Coville, U. S. Department of Agriculture.

In our last lesson we studied a plant which we cultivate in our gardens on account of its beautiful flowers; to-day we will take up several plants which we cultivate on account of the food they furnish either for ourselves or for domestic animals.

Of some plants, as of beans and peas, we eat the seeds; of the melon, the cucumber, and the tomato we eat the fleshy covering of the seeds. The seeds of many plants are a very valuable food material, because they can be preserved for an almost indefinite length of time, provided they are kept dry. How can we keep the fleshy parts of fruits and vegetables?

If you closely examine a head of cauliflower, you will find that it consists of thickened flower stalks and undeveloped flowers. Can you tell now which parts of the cauliflower we eat?

Of the common lettuce we eat the green leaves, which we prepare in different ways. Can you tell now what parts of cabbage, kale, and spinach appear on our tables? How many wild plants do you know that are eaten as greens?

Of the kohlrabi we eat the enlarged succulent stem, and of white or bleached celery the tender leafstalks are much relished by some people.

Turnips, carrots, and rutabagas, and the different kinds of beets are biennials, which means that they live two seasons and do not produce seeds before the close of the second season. They store much food material in their large roots,

Return of the blackbirds, wild ducks, and geese, and other birds.

and it is on account of the value of their roots that we cultivate them. Can you mention a root which we frequently
use as an appetizer? It is a root which can easily bring
the tears to your eyes.

After this lesson has been given, let the pupils plant some
beans, peas, cucumbers, a few roots, and some corn and
wheat, and let them carefully observe the life cycle of these
plants. Pay special attention to the germination of the
different seeds.

§ 4. The Apple Tree and Other Fruit Trees.

MATERIAL: Twigs of apple tree, plum, and cherry, with flowers
and leaves ; tulip flowers for comparison.   Before the lesson is given,
children must have observed how bees and other insects gather honey
and flower dust on the blooming trees.

Who of us has seen anything more beautiful than an
orchard in bloom? The trees look as if they were covered
with snow, sweet fragrance of countless flowers meets us
on the gentle breeze, we hear the buzzing of swarms
of busy bees, and watch them delve into flower after
flower.

Let us now compare the white or pink flowers of the
apple tree with the flowers of the tulip. We find five
beautiful flower leaves, which together form the crown or
corolla of the flower. Can you tell why we might call them
the crown? Under the corolla leaves we find five small
green leaflets. These are called the cup, because in many
flowers they are grown together and form a small cup.
Within the corolla leaves we find a great many little dust
threads, and in the centre is a tiny pistil with a scar on
top. How do the tulip and apple blossoms differ?

Besides the flowers we find green leaves on the twigs.
These leaves have a saw-toothed margin. How do these

Observations. — Appearance of blow-flies, mosquitoes, and other insects ;
return of the swallows.

leaves differ from tulip leaves? How do these twigs differ from the stem of the tulip?

The twigs and branches of the apple tree generally form a large bushy top, which is supported by a short, stout trunk. Why do you think the trunk must be so strong? What other fruit trees do we often find in an orchard?

Bees and other insects come to the flowers to get the honey out of them. Try to find the little drops of honey. While they are collecting honey, they often carry some of the dust of one blossom and accidentally brush it off on the scar of another blossom. (The children must see this and find the flower dust on the bees.) Do insects ever eat the ripe apples?

§ 5. The apple tree is really the fruit tree of the northern half of our temperate zone. In most of the Middle States large quantities of apples are produced every year, while only a few varieties will do well in some of the Northwestern States. Apples will keep for a longer time than most other fruits, and can therefore be shipped to great distances. They are also sold dried and canned. The juice of apples may be pressed out and made into cider. We must watch the fruit growing after the blossoms have faded.

In some parts of our country wild apple trees grow in the woods; but their fruit is very much smaller than that of the cultivated trees.

In order to make sure that a fruit tree shall bear choice fruit, the young tree is cut off slantingly when it is only a few feet high. Now a twig of the desired variety of fruit is cut in the same way and securely tied with its cut surface upon the cut surface of the young tree. If the operation is successful, the twig and tree will grow together and later the tree will bear the kind of fruit which the twig

Keep some tadpoles in your aquarium and watch their growth.

would have borne if it had remained on its own tree.   This
process of improving the fruit of trees is called grafting.
If only a bud is transferred from one tree to another, it is
called budding.   There is quite a number of methods of
grafting besides the one described.   In a tree thus improved, the roots of the wild stem furnish the moisture and
mineral food, while the transferred twig develops into the
flowering and fruit-bearing top.

§ 6. The Apple-tree Tent Caterpillar.   *Clisiocampa Ameri-*
cana.

MATERIAL : Eggs and caterpillars with tent ; May beetles, or June
bugs and their larvæ, if procurable ; several twigs with leaf-lice on
them.   Try to find some ladybugs and their larvæ feeding on leaf-
lice.   Keep some of the caterpillars in your observation box ; have the
children feed them, and observe their development.   Describe the
moth, if you get any.

Gardeners and farmers have to fight a great many insect
pests.   Some of these insects lived on wild plants, before
our country was settled, but, like the potato beetle, have
now taken to eating cultivated plants.   Others, like the
white cabbage butterfly, have been accidentally brought over
from Europe.

To-day we shall study the tent caterpillar, which has
spread over nearly the whole of the United States.

In the month of July a small, reddish-brown moth lays
about three hundred eggs, which are fastened in the shape
of a ring around the small twigs of fruit trees.   The eggs are
firmly glued together and covered with a kind of varnish.
The next spring, about the time the leaves begin to grow
on the trees, the caterpillars hatch, make their first meal
on the glue with which the eggs are covered, and then
they begin to eat the tender young leaves, and to weave a
little tent.   Into this tent they crawl towards evening

Observations. — Flowers of maples, box elder, and pines.

and remain in it during cold and stormy weather. As the caterpillars grow, they make their house larger. They have a very good appetite, and if they happen to live on a small tree, they sometimes strip it entirely of its leaves.

In about five or six weeks they are full grown. Describe some full-grown specimens. Note their size, color, hairs, legs, etc. Now they leave the trees and hide separately in

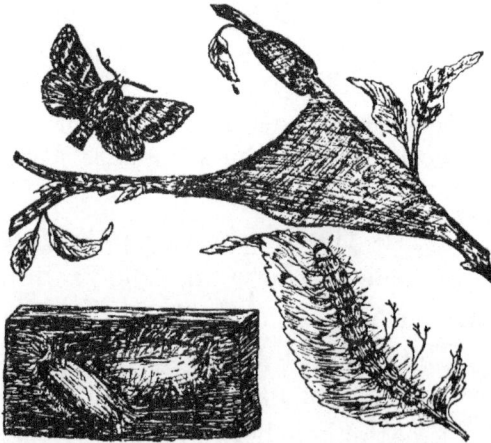

FIG. 1. APPLE-TREE TENT CATERPILLAR.
Eggs, tent, caterpillar, cocoons, and moth. All reduced. After Comstock.

some sheltered place, as under the cap-boards of fences. Here they spin a cocoon of yellow silk, and in this the caterpillar changes into a chrysalis.

Within two or three weeks the chrysalides change to moths, which break through the cocoon and escape. The moths do not eat anything; they lay their eggs and then die, after they have lived only a few days. Have you ever seen butterflies eating anything? What does the chrysalis eat? You may find these caterpillars on various trees, but especially on the apple tree, the wild cherry, and the wild plums. There are other species of tent caterpillar which

Let the children plant beans, cucumbers, etc., outdoors or in pots and boxes.

live on different forest trees. Find some other caterpillars
and feed them in an observation box.

Remedies. — The tent is easily seen and should be destroyed early in the morning or after sunset. The caterpillars seldom leave the nest before 9 A.M. A sharp, practised
eye can also easily find the egg clusters in winter.

NOTE TO TEACHER. — If you cannot procure this insect, you might
take the white cabbage butterfly instead.

§ 7. Plant Lice or Aphids.

MATERIAL : Twigs of apple, cherry and other trees infested by
plant lice ; different kinds of ladybugs and their larvæ feeding on the
plant lice.

Almost as soon as the leaves of apple and cherry trees
are out, we find some of the young shoots infested by very
small, yellowish-green or dark insects. Sometimes they
literally cover shoots and leaves. They have a small sucking beak like a mosquito, and injure the plants by sucking
their juice. The attacked leaves generally become twisted
or curled, and thus even furnish shelter for their enemies.
These little pests are found on very many wild and cultivated plants. You may have noticed some peculiar swell-
ings on the leafstalks of cottonwood trees. If you open
these bulbs, you will find them full of plant lice.

In early spring the aphids are wingless, but later in the
season there are some with wings, and these fly to other
plants and become the founders of new colonies. Late in
fall eggs are laid on the buds of food plants. These eggs
can endure the frost, and hatch the following spring.

Most aphids secrete a sweet fluid, by which bees, flies,
and ants are attracted. You may sometimes find the leaves
of box elders and other trees almost covered with this

Observations. — Pay special attention to germination and to the development of the seedlings.

honey; thousands of bees and flies then buzz about such trees. You have no doubt seen ants on the colonies of aphids. Ants will eat almost any insect they can kill, but to the aphids they go to get honey; they even keep some of them over winter in their burrows.

Remedies. — If trees and shrubs are sprayed with strong soap-suds, weak lye, or tobacco water, as soon as the lice begin to appear, very many of them will be killed. They cannot be made to eat Paris green. Why not? If they appear on house plants, the latter may be fumigated in the following manner: Put the plants into some closet or room which can be closed, then burn some moist tobacco or tobacco refuse in the closed room or closet. If fumigated too long or too strongly, the plants are apt to be injured.

NOTE TO TEACHER. — For a detailed description about remedies for insect pests, diseases of plants and animals, noxious weeds, etc., consult the reports of the Agricultural Experiment Station of your own state and the Farmers' Bulletins, published by the Department of Agriculture, Washington, D. C. These reports and bulletins con tain much valuable and practical information, and every teacher, gardener, and farmer should send for a list of them and procure those which interest him. They are sent free to any one who applies for them.

§ 8. Some Insects that are Beneficial to Man.

MATERIAL: Ladybugs as in § 7.

If insects had no enemies, which eat them and their eggs, and if the weather was always favorable for them, they would soon become so numerous that they would kill all the plants on which they feed.

You know that many birds eat insects, but one insect often eats another. Certain kinds of small flies lay their eggs on caterpillars, and when the little maggots hatch, they eat themselves into the caterpillar, make him sickly, and

Watch the growth of blue flags near ponds and lakes.

often cause his death.   If he lives long enough to change
into a chrysalis, the little maggots still live in him and
finally kill him, before he can change into a butterfly.   The
little maggots then remain in the chrysalis until they have
transformed themselves into flies.   These little flies are
either Ichneumon Flies or Tachina Flies.   Nearly every
caterpillar has one or more of these little flies as his special
enemies.   If you open a considerable number of chrysalides,
you are very apt to find the little flies or their cocoons in them.

Another class of very useful insects are the Ladybirds or
Ladybugs.   You will often find them and their larvæ on
plant lice, which they devour by the dozen.

FIG. 2.   LADYBUG AND LARVÆ ON ROSE LEAF.

Besides the insects
eaten by birds and de-
stroyed by other insects,
a great many are killed
by frost and by cold,
rainy weather.   Some-
times they starve in
great numbers, because
their food gives out.

What becomes of the millions of insects when cold weather
sets in?   Where do they come from in spring?   Try to
answer these questions for a few insects by your own
observation.

NOTE TO TEACHER. — For a more detailed account of the insects
just studied, see

Comstock.   Manual for the Study of Insects.
Harris.   Insects Injurious to Vegetation.
Saunders.   Insects Injurious to Fruits.

For injurious insects in your neighborhood, see your State Reports, or
write to your state entomologist, sending specimens with your in-
quiry.   See also, A Study in Insect Parasitism, by L. O. Howard,
Department of Agriculture, Washington, D. C.

§ **9. The Robin.** *Merula migratoria.*

MATERIAL : If possible, mounted robin or colored picture ; nest of previous season, but well preserved. Pupils must have observed outdoors : Arrival of robins, building of nests, song, and food.

NOTE TO TEACHER. — For your own information about birds, consult :

Mabel O. Wright. Birdcraft : A field-book of two hundred Song, Game, and Water Birds.
Grant. Our Common Birds, and How to Know Them.

The report on the birds of your state, if one has been published.

Who of us has not been happy, when, in the latter part of March, he saw a robin perched on the top of a tree and was charmed by its cheerful song, which seemed to assure him that spring has actually arrived ? As the harbinger of spring, we all love the robin.

From the tip of his bill to the end of his tail he measures about ten inches. His yellow bill contrasts strongly with the black head and olive-gray back. His tail is black, but we know him best by his brick-red breast, which has earned him the name of Robin Redbreast. If you have sharp eyes, you will detect that in the fall our robin assumes a paler garb, as very many birds do. The female robin is always paler in color than the male. Can you distinguish male and female of the Rose-breasted Grosbeak and the Scarlet Tanager ? Is it of any advantage to female birds not to have as bright a plumage as the males ? Compare the color of males and females among our domestic fowls.

Very early in the morning, before sunrise, the robin begins his song, together with many other birds. He continues to sing until about the middle of July. Mrs. Wright has translated his song into words. She hears him say, " Cheerily, cheerily, cheer up, cheer up ! " or, " Do you think what you do, do you think what you do, do you thi-n-k ? " How would you put his song into words ?

If he is not molested by cats and bad boys, the robin seems to be fond of man's society. I have found his nest within ten feet of the house, where a farmer and members of his family worked and passed to and fro all day long. I noticed, however, that he tried to leave and approach the nest unseen. Where is his nest usually placed, and of what is it made? What sound does the robin make when you alarm or frighten him? How many eggs are laid in the nest? Of what color are they?

The robin feeds on all kinds of worms and insects. Watch him running along on the grass catching crawling and flying insects and sometimes pulling a worm out of the ground. It is true that he, like his cousins the Catbird and Brown Thrush, is very fond of berries. But should we not be willing to pay gladly this trifle for his music and his company? Once, when a late fall of snow had caught many of our summer birds, I saw a robin greedily eat crumbs of bread thrown on a cleared place.

You must watch a pair of robins from the time they begin to build their nest until the young leave the nest. Make careful notes of your observations and afterwards connect them into the form of a composition.

§ 10. The Garden Rose.

MATERIAL : Twigs and flowers of different kinds of garden roses and wild roses ; fresh hips and hips of last season ; twigs of wild plum with thorns. The children should have observed the breaking of the buds and the opening of the flowers, before the lesson.

If we were asked which one of all the flowers we think most beautiful, most of us, I think, would vote for the Rose as the Queen of Flowers. Here we have them large and small, deep red, snow-white, blushing pink, and yellow. Even in the wild roses of the prairie and the woods, we find an unlimited variety of most delicate hues.

Before the delicate flower opens, it is enclosed by the green cup or calyx, which ends in five leaflets. These are bent back in the open flower. Within the calyx the wild roses have five corolla leaves or petals, while the cultivated species have a large number of petals. Inside the petals we find a ring of dust threads, carrying the yellow dust bags. The dust threads are generally called stamens, and the little bags are called anthers. In the cultivated roses many of the stamens with their anthers have been changed to petals. Where are the stamens inserted? In the centre of the flower you see a bundle of tiny pistils. Are these more distinct in the wild, or in the cultivated kinds?

After the beautiful flower has faded and the delicate petals have dropped to the ground, the globular, or pear-shaped calyx of the wild roses continues to grow, and within it grow the hard seeds. We shall examine some of these hips later in the season. Late in fall, when frost has killed flowers and leaves, these red hips still adorn the wild bushes. Many cultivated roses bear no seeds. How could you propagate them?

Even the leaves of the rose are beautiful. The whole leaf consists of five or more leaflets, and each leaflet has a nicely saw-toothed margin. A leaf consisting of two or more leaflets is called a compound leaf. The leaves of the rose and of clover are compound; those of the tulip are simple. Compare the leaves of a number of plants, when you stroll through the woods or over the prairie. Did you find any insects which seem to make the leaves and flowers of the rose their pasture? Are they beneficial or injurious to it?

The stem of the rose is woody and lives several years. Some roses are low shrubs, others are quite tall, and some climb on trees or walls. You have all heard the proverb: "No rose without thorns." Botanists call these thorns prickles. Do they grow out of the wood or from the bark? Here, on the twigs of the plum and the hawthorn, you can see some real thorns, or spines. See if you can show that they are connected with the wood and are really stunted branches. Of what use might prickles and thorns be to plants? Examine the stems of young trees and shrubs in any thicket in early spring, and you will find that the rabbits have peeled off the bark of many species; that they have cut off and eaten the smooth branchlets of roses, but that they have touched the prickly stems of roses and green brier very sparingly. Can you find evidence that other animals besides rabbits have injured shrubs and young trees?

All our wild roses are hardy shrubs well able to winter in the regions where they grow. They increase by seeds and also by rootstocks. Many garden roses need some protection during our Northern winters. Gardeners generally raise the cultivated roses from slips or cuttings. They also employ grafting and budding on roses and other ornamental shrubs as well as on fruit trees.

§ 11. Review and Summary.

We have observed that some flowers do not grow wild, but grow only in our gardens, where we take care of them. If we did not sow or plant them every spring, most of them would soon disappear. Do you not think that many of our wild flowers are so beautiful that they deserve a place in our gardens? In Europe, our Goldenrods, the Butterfly Weed (a milkweed, *Asclepias tuberosa*), the Sumach, and other flowers and shrubs are cultivated in gardens.

Cultivation has brought about many changes in plants. The fruits have become larger and sweeter. In some, as the seedless orange, the seeds have entirely disappeared. In many plants, the flowers have become double and larger, but are often sterile. What change has been effected in the cabbage, cauliflower, and kohlrabi? What becomes of the edible parts of these plants if they are left in the garden over winter? Do these fleshy parts begin to grow again if you winter the plants in the cellar and plant them out next spring? When do the vegetables you have studied produce flowers and seeds?

If possible, compare the flowers of turnips, radishes, rutabagas, cabbage, mustard, cauliflower, and kohlrabi. Do you find any remarkable similarity?

Many unbidden guests help themselves to our fruits and vegetables or eat the leaves of our plants. Among them are caterpillars and beetles. The Forest Tent Caterpillars sometimes become so numerous that whole acres of woods are stripped of leaves and many trees killed. If fruit trees are denuded of their leaves, they cannot bear fruit, but must produce new leaves or die.

In fighting our insect enemies, many birds, and even some insects, come to our assistance. It is therefore to our interest to protect these birds and insects as much as we can.

Many insects come to the flowers for honey, but incidentally they carry the pollen of one flower to the pistil of another. This transfer of pollen from one flower to another is called 'cross-fertilization. As many flowers are so constructed that they cannot fertilize themselves, but depend entirely upon insects for fertilization, such insects render a great service to the plant. If the pistil of a plant does not receive pollen in some way, then that plant will not produce any seeds.

EXPLANATION OF SOME BOTANICAL TERMS: —

1. The Calyx, The cup. — It forms the green covering of the flower bud, and is found under the corolla, after the flower has expanded.

2. The Corolla, The crown. — The bright-colored leaves or the tube within the calyx.

3. The Stamens. — The dust threads.

4. The Pistil. — The part or parts of a flower which contain the rudimentary seeds. Pistils are the central organs in common flowers.

5. Sepals. — Parts or lobes of the calyx.

6. Petals. — Parts or lobes of the corolla.

7. Anthers. — The dust bags.

8. Stigmas. — The scars or lobes of the pistil.

II

POND, LAKE, AND STREAM. MAY TO JULY

NOTE TO TEACHER. — A walk to a place, where, at least, several of the features mentioned below can be seen, must precede this lesson.

MATERIAL: Twigs of willows, alder, tamarack, dogwood, rushes, wild rice, sagittaria, water lilies, and other aquatic plants; various aquatic insects. Place some stagnant water in a glass and notice the large number of very small animals in it. The children should have observed the swimming and diving of ducks and geese.

§ **12.** Near standing and running water, we find the plants entirely different from those on high ground. Poplars grow on the moist slope and willows frequently form thickets in swamps and near streams and lakes. Other characteristic plants are the tall grasses and rushes, which fringe our lakes and rivers and often entirely cover marshes and ponds. All of you know the beautiful cat-tails, of which the boys sometimes make torches by dipping them into kerosene, and you have also observed the tall, round rush, which grows in streaks and patches in our lakes, where the water is from two to six feet deep. Where the lake or pond has a muddy bottom, you found the lovely water lilies with their large leaves floating on the surface. You also found, in shallow places, very small roundish green leaves with very short rootlets, that did not reach down into the mud. These little plants are called duckweeds. Along the shore we found many frogs.

Observations. — Along roads, streets, railroads, on ploughed fields, and near streams, find places where running water has washed some of the soil away.

When they saw us approaching, they jumped into the water, but turned right back to shore. I wonder whether the boys can tell us why the frogs were sitting along the shore. Why do they not go far into the lake or river? In quiet shallow places we noticed several kinds of small animals wriggling or darting about. The minnows and young fish live on these.

If we can hide near the shore of a large pond or lake, we may see some wild ducks alight on the water. They swim about very gracefully and search for food, which consists of small fishes, insects, roots of water plants, and wild rice.

Have you ever seen the Great Blue Heron fishing? I believe he catches more fish than many a small boy.

The Blackbirds and Bobolinks made the bushes and reeds lively with their song, but when we approached their nests, they gave us a sound scolding and seemed very much inclined to fight us. Why are the mosquitoes so abundant about lakes and streams?

§ 13. Willows and Poplars.

MATERIAL: Twigs and fruit catkins of different willows and poplars; flower catkins, if still to be found.

NOTE TO TEACHER. — It is not necessary to give any specific names of willows to the children. The family is easily recognized by the flowers; but it is often difficult to distinguish the numerous species.

Very soon after the snow had melted, we found the pretty pussy willows. When we examined them closely, we were a little surprised to find two kinds, but only one kind on the same bush or tree. The catkins on some bushes consisted entirely of tufts of stamens with large yellow or reddish anthers, while on others they consisted of small greenish ovaries, with little scars or stigmas on top. Honeybees,

---

Observations. — Try to find the places where some of this soil was again deposited.

bumblebees, and other insects carried the dust or pollen from the sterile, or staminate flowers to the fertile, or pistillate ones. Why did the bees visit the willows?

After the insects had fertilized the pistils, the seeds in them grew, and many of the little pods are opening and the tiny seeds which are attached to a tuft of delicate cotton are carried off by the wind and scattered far and wide. If they happen to drop on moist soil, the young trees soon begin to grow.

From the twigs and leaves we have gathered, we see that there are many species of willows. Most of them have rather long and pointed leaves; some have the leaf margins toothed, in others they are entire. The lithe branches of some

FIG. 3. SAGE WILLOW. *Salix candida.*
a, staminate catkins; *b,* pistillate catkins.
About one-half natural size.

species are well adapted for all kinds of wickerwork. Your mothers' work-baskets are made of willow twigs. On the prairies they are often planted for wind-breaks. Most of them like to grow near the water, but several do quite well on high ground. Some are small bushes and shrubs; others are trees fifty feet high. About twenty species of the willow family grow wild between the Atlantic coast and the Rocky Mountains, and quite a number have been intro-

Which was carried farther, the coarse or the fine material?

duced from Europe. Willows grow not only from seeds,
but also from the roots and from cuttings. Try to grow
them from cuttings. It is easy to cultivate them, but hard
to kill them. Why are they well adapted for wind-breaks ?
Have you ever seen the caterpillar-like larvæ of the large
sawfly on them ? (See Minnesota Entomologist's Report
for 1895.)

§ **14**. Closely related to the willows are the poplars.
Their flower catkins also appeared very early, before or with

FIG. 4. QUAKING ASP. *Populus tremuloides.*
a, staminate catkins; *b*, pistillate catkins. About one-half natural size.

the leaves. We found the sterile and fertile catkins on
different trees, the same as on willows, but the poplar cat-

Observations. — Visit a stone quarry and a brick-yard.

kins were longer and were hanging down. There was no honey in them and therefore no insects were attracted to them; but the least breeze shook the long tassels and car-ried the pollen from the staminate to the pistillate flowers. The poplars are fertilized by the wind. Can you think of other plants that are wind-fertilized?

The largest poplar is the Cottonwood Poplar. In rich and moist soil it grows very fast, and sometimes reaches a height of one hundred and fifty feet. The seeds of the poplars resemble those of the willows, and they are also disseminated by the wind. Other common and well-known poplars are the American Aspen and the Balsam Poplar, or Balm of Gilead.

Which poplars are planted for shade and ornament? Do you know what their wood is used for? Can you see any reason why poplars and willows should be much more common than butternut and walnut trees?

§ 15. The Frog.

MATERIAL: In early spring, place some frog eggs in a fruit jar. Change the water about every third day. Take the water out of a brook or pond where frogs and tadpoles live. Do not take well or cistern water. Feed the tadpoles with a few crumbs of wheat bread or water crackers. Later on, put a few small water plants into the jar; change them when no longer fresh. When the legs begin to grow on the tadpoles, place a few pieces of wood or a small ladder into the jar. Several species of frogs in a fruit jar. The jumping, swimming, and croaking of frogs must have been previously observed outdoors. You ought to have some eggs and tadpoles at different stages of de-velopment preserved in weak alcohol.

In early spring we found clumps of frog eggs. Each egg looked like a little ball of jelly, and had a dark spot in the centre. The gelatinous covering at first protected the real egg in the centre, but when the little tadpole began to develop it fed on this jelly. What did it eat later on?

Is the rock in the quarry stratified?

The head of the young tadpole is not set off from the body, which fact gives the little wiggler the appearance of being simply head and tail. A margin of skin is attached to the upper and to the lower side of the tail and, acting like a fin, enables the animal to swim about freely. On both sides of the head we see little appendages. These are the gills, and with them the animal can breathe the air which is in water. But while the body grows, the gills shrink and soon disappear. Now the animal breathes through lungs, which have grown, while the gills were shrinking. But we noticed some more wonderful changes. A pair of hind legs

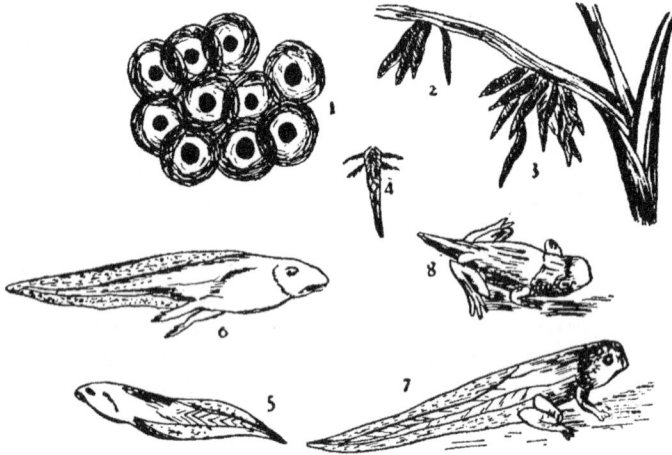

FIG. 5. FROG EGGS AND TADPOLES.
1, frog eggs; 2–4, very young tadpoles; 5–8, older tadpoles. All except 1 reduced.

began to grow, the tail shrunk, then the front legs appeared as little stumps, and finally when the tail had entirely disappeared, the funny little pollywog had changed into a cute little frog.

The change in its organs enables the animal to lead a different life.

Observations. — What differences do you notice between clay and sand?

The tadpole could not live in the air on land, because it had no lungs, but the frog can; he can also swim and dive, but comes to the surface to breathe.

How tadpoles and frogs feed. — The tadpole in the pond eats whatever small creatures come in its way and whatever decaying matter it can find; but Mr. Frog is a hunter. It is true he does not run his game down; being a gentleman inclined to take things easy, he sits down and waits for the game to come within reach. Once I watched a large green frog catching his breakfast. He was squatting on the moist ground near a brook, when two large butterflies (some species of the Fritillaries) came sailing along to quench their thirst. The frog's large eyes rolled and seemed to fairly bulge out of his head, and he raised himself up to watch them better. One of the butterflies alighted within about six inches of Mr. Greencoat. The frog crouched, crawled up behind a little elevation to within about three inches of the butterfly; then he sprang like a cat, his tongue caught the insect, a few quick movements of his jaws, and the butterfly, wings and all, had disappeared. If you have patience, you may observe frogs catching all kinds of insects. On another occasion I found two large June-bugs in a frog's stomach. The frog's tongue is attached in front and not behind. In catching an insect it tips over, the insect sticks to its slimy surface and is drawn into the large mouth.

We have seen how tadpoles and frogs are well adapted to find their food each in their own way and in their peculiar condition.

We shall now try to find out what means the frog has to escape from his enemies or defend himself against them. Ducks and other water birds eat frog eggs whenever they can find them. Fishes, snakes, and birds devour thousands of young tadpoles, although the little fellows try their best

Look for weeds along roads, in fields and gardens, and in waste places.

to hide in the mud. Sometimes the ponds dry up and the tadpoles die for want of water. But so many eggs are laid that there are always a great many that escape all perils and develop into the frogs which are better fitted to hold their own in the great struggle in nature. Their life is, indeed, not without trials. Small mammals, many of the larger birds, snakes, bass, and pickerel are always hunting for frog steak. Scarcely is the frog music heard in ponds and pools, when from every town in the land an army of small boys with murderous-looking spears takes the field against poor froggy ; every pool within walking distance of the youngsters is invaded, and "Frog Legs, 25¢," appears on the bill of fare in every restaurant. This warfare against them in the breeding season has thinned out the greencoats around our cities. A little later in the season fishermen use thousands of frogs for bait, and when fall comes, schools and colleges demand them for purposes of study

Thus we see that his miseries never end, his enemies are always with him. And yet he survives. How does he ? How can he ? Let us see.

His brown or green spotted skin blends perfectly with his surroundings. As long as he sits still, you can hardly discover him among grasses, leaves, and sticks. He sees and hears well, although he has no outer ears. A delicate membrane behind the eyes covers the ear opening. As soon as he perceives any possible enemy coming too close, he bounds into the water, where he conceals himself in the mud or under plants. If water is not near, he hurries towards tall grass, weeds, or brush. How are his legs fit for jumping? How for swimming? His favorite territory is the borderland between high land and deep water. Although he is a fine swimmer, he always stays near the shore. Why? You have seen how the frog eludes and escapes his enemies. Has he any means to fight them?

How frogs hibernate in winter. — You have noticed that in late fall, when insects disappear, most birds also leave us and go south. The frogs also disappear at this time. Where do they go? They hide in the muddy bottoms of pools, lakes, springs, etc. Here at a depth where the frost of winter will not touch them they sleep a death-like sleep. They eat nothing and do not breathe during the whole winter. *This kind of sleep is called hibernation.* When the sun and the warm south wind have melted the ice from our waters, the frogs awake and come to the surface. The males begin to call their mates by croaking, and soon the females deposit their eggs.

There is quite a number of different kinds of frogs in our country. Closely related to the common frogs are the Toads and Tree Frogs or Tree Toads. Toads live mostly on land, Tree Toads on trees, but both deposit their eggs in water. *Frogs and toads are very useful animals, because they eat a large number of injurious insects.*

§ 16. The Mosquito.

MATERIAL: Mosquitoes, eggs, and larvæ, if possible. The larvæ, often called wrigglers, can be found in almost any stagnant pool and in rain-water barrels. Previously observed: How mosquitoes fill themselves with blood; how horses and cattle try to rid themselves of the pest.

Perhaps you have made out most of the life history of the mosquitoes from the observations you have made in your little aquariums. The eggs are laid in little rafts, which float on the water. In a few days the lower ends of the eggs open and the larvæ escape into the water. Each one of them has a little tube near the tail end. By means of this tube they breathe. When they are at rest, they hang with their heads down in the water and have their breathing tubes at the surface in contact with the air. The larva lives on decaying matter in the water; it grows very fast, and throws off its skin several times. *The process of shedding the skin is called moulting.* Because these larvæ swim with a wriggling motion, we call them wrigglers or wigglers.

The wrigglers change into club-shaped pupæ. This pupa has two breathing tubes attached to its thorax. It swims by means of two little appendages on its tail end. After a few days the skin of the pupa splits open on the back, and the winged mosquito carefully works out of it. It rests on

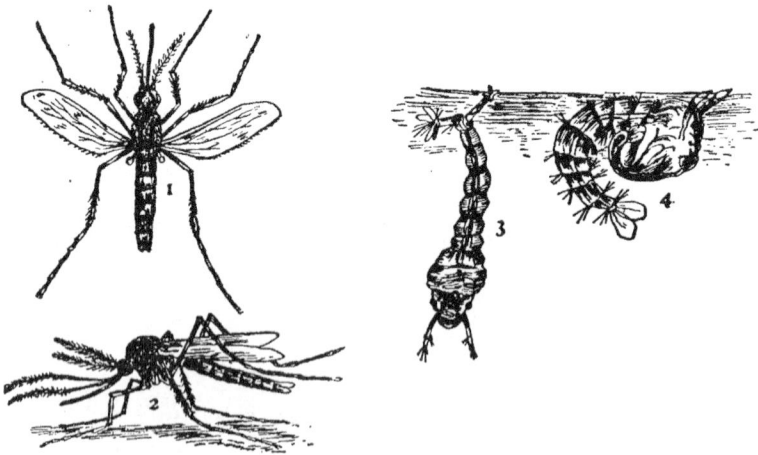

Fig. 6.   1 and 2, mosquitoes; 3, larva; 4, pupa.   All enlarged.

the skin until its wings have hardened a little, then it flies away.

Now after we know the life history of this little pest, we may think about how to fight them.

It is plain that we should not breed them in barrels or stagnant pools near our homes. In small pools the larvæ can be killed by a little kerosene poured on the water. The oil prevents them from breathing.

Smudge fires offer some protection against the mosquitoes. Hunters and woodsmen often anoint their faces and hands with a mixture of mutton tallow, camphor, and oil of pennyroyal. Different other mixtures also keep them off. It is only the female mosquitoes that bite us, or keep us awake by their music. The males can neither bite nor sing.

Can you tell why mosquitoes are more common in a wet summer than in a dry summer? Why are there so many in most wild districts and comparatively few in old settlements and in cities? Why are there so many near the water? What birds have you seen eating them? Have you seen the large Dragon Flies eat them?

§ 17. The Swallows.

MATERIAL: Pictures of different swallows or mounted birds. Previously observed: Flying of swallows about running and standing water; how they gather mud for nest building; how they build their nests and feed their young.

When mosquitoes and flies appear about our streams and ponds, the swallows arrive too. They travel by night or in the early morning and some fine spring morning, when we look out of our window, they are there. All day long they are on the wing, circling about near creeks and ponds, among cattle and horses, catching insects in the air. *Their long pointed wings, worked by strong muscles, never seem to tire. But their feet are small and weak, because they make little use of them.* When and where have you seen them sit down or perch? Even the stingiest farmer can find no fault with these lithe denizens of the air; for not even a single berry do they ever poach from his garden, so exclusively do they live on insects.

There are three species of swallows very common through the United States and Canada, wherever local conditions are favorable. The best place to observe them is about their nests.

The smallest, the little mouse-colored Bank Swallow, or Sand Martin, lives in holes excavated into high river banks, in sand rock, or in other convenient places. They are often found in large colonies.

The Barn Swallow has a beautifully steel-blue back, *a deeply forked tail,* is rich brown below with an almost brick-red throat. Its nest looks like a quarter of a hollow sphere, and is attached to rafters and beams in barns and stables.

FIG. 7. BARN SWALLOW.

The Cliff Swallow generally builds a gourd-shaped nest with an opening in the neck. In settled districts, the nests are often placed in rows under the eaves of barns. In wild districts the birds are compelled to build under cliffs which formed their original homes. This swallow is also steel-blue above, but dusky white below, with a white, crescent-like frontlet.

Watch either the Cliff or Barn Swallow build its nest. Tell how they do it and what material they use. After the eggs are hatched, watch the parents feed the young. Do both parents feed them? How often do they return within an hour? Are the swallows as wild as the Great Blue Heron? Do they seem to love the neighborhood of man? Why?

Early in August the swallows begin preparations for leaving this part of the country. They assemble in large flocks; at night they roost on trees. Willow thickets near lakes and streams seem to be favorite places. In the summer of 1895 I found thousands of them roosting in a willow thicket on a small island in the Des Moines River near Windom, Minn. Where have you seen them assemble in the daytime? About August 20 they have generally left. Why should they leave so early? The herons do not leave until the middle of October, and wild ducks are often here as late as November. Why?

Do you know the Purple Martin and the Tree Swallow? Do you know on what the Chimney Swift and the Night Hawk live? How do they catch their food?

§ 18. The Blackbirds.

MATERIAL: Picture of bobolink or mounted bird; similar material for different species of blackbirds; nest of some blackbirds showing how it is fastened to reeds or twigs; some warbler's or sparrow's nest with one or two cowbird eggs in it.

Who is not familiar with the Blackbirds? A merry tribe they are indeed; flying about, singing, scolding, wooing the whole summer day! Everybody knows the large Purple Grackle, or Crow Blackbird; the Red-winged, and the Yellow-headed Blackbird. In the nesting season you will generally find their wives, clad in modest gray and brown, near them, both males and females ready to scold and fight any intruder of their reeds and rushes. *But the favorite in the whole family is certainly the Bobolink.* It is found in this region from May until September. The head of the male

FIG. 8. BOBOLINK.

is black, a patch of light brown marks the nape; but most conspicuous are the white rump and the white wing coverts. The rest of the body is black, with the exception of the brown bill and a little buff edging on some tail and some wing feathers. The female is yellowish-brown below and is striped with brown above, with yellow and white tips to some feathers on the rump. After the young are hatched, the male doffs his clownish dress and appears in as sombre a garb as the female.

The nest is built in a tussock on low meadows; it consists mostly of dried grass, and is very difficult to find. If two persons drag a long rope over a piece of meadow where bobolinks abound and carefully note where the birds fly up, they may succeed in finding it. It contains from four to six brownish or grayish eggs which are marked with spots or blotches of darker brown.

The song is really indescribable, but is very characteristic of the bird. You may see a bird nervously trying to sing his best on some knoll; suddenly he rises to a considerable height, drops down again into the grass just as suddenly, all the time bubbling over with joyous melody. If you find the right meadow, you may see and hear a great many bobolinks at the same time.

Later in the season the bobolinks are all plain brownish-striped birds. They linger for a while, rambling about fields and meadows, and are sometimes seen in flocks on oats, of which they are fond. About the 1st of September they begin to migrate south. They feed now mostly on wild rice, and are known as reedbirds or ricebirds to gunners, and thousands are killed for the market. Do you think we ought to shoot these beautiful little singers for the table, when we have so many large game birds?

When Blackbirds become very numerous in fields, they will do some injury to grain; but early in the season they live largely on insects. It must be admitted, however, that the bobolinks are very injurious to the rice-growers in the South. Thousands of these ricebirds hover constantly about the rice fields, and the Southern farmers have to spend thousands of dollars annually to keep the birds from eating all the rice they can grow. Although boys and men hunt and shoot them daily, their number has not diminished.

NOTE TO TEACHER. — Yourself and pupils interested should send for a copy of Some Common Birds in their Relation to Agriculture, by F. E. L. Beal, U. S. Department of Agriculture, Farmers' Bulletin, No. 54. This valuable pamphlet gives information about the food of more than twenty of our common birds, and contains very good figures of as many birds.

§ 19. The Great Blue Heron.

MATERIAL : Picture of great blue heron and other herons. If possible, pupils should have observed the great blue heron in his native haunts. The teacher ought to have made these observations, if the pupils could not. Visit a heronry if you can. Draw a life-size picture of the heron on the blackboard. Substitute some other member of the heron family, if you prefer.

How many of you boys have ever been hunting ducks in September or have rambled along lakes and streams in summer? You have found secluded bays, which with high trees around them, with tall thin rushes growing in patches in the placid water and gently swayed by the summer breeze, and with water lilies peeping out from amongst the grand, floating leaves, make a most charming picture. If you are lucky enough to see one or two Great Blue Herons fishing in the shallow water or frogging along the shore, then you have indeed a landscape with a soul, with real, visible life.

This magnificent bird stands fully three feet high. On account of its slate-blue back and large size, it is called the great blue heron, or blue crane. He is the monarch of the lake shore and the terror of frogs and fish. Just as patiently as the frog waits for insects to approach him, the blue heron waits for fish to come within reach of his long sharp bill. Or he walks gravely along the water's edge and spears the frogs he happens to hunt up. His long legs, neck, and bill excellently fit him for this kind of hunt. He sometimes catches a fish ten inches long. If he sees you watching him, he will spread his wings, slowly rise, and fly to some undisturbed spot. How does his flight compare with that of the teal ducks?

A great many herons build their nests close together. These colonies are called heronries. They are generally located on an island in a lake, or in a river bottom. The best-known heronry in Minnesota is on Crane Island, in Lake Minnetonka, about fifteen miles from Minneapolis. On this island hundreds of herons have built their nests on the large trees for so many summers that even the Indians could not tell when the herons first came there. If your boat passes the island, you see the great blue herons and the black-crowned night herons perched on the trees along the shore; here and there the large nests appear as bundles of sticks. Minnetonka is now the most popular summer resort in Minnesota; large and small steamers, any number of sailboats, and hundreds of rowboats move about on the lake all summer; still the herons and the double-crested cormorants continue to inhabit Crane Island. It was, however, found necessary to make them wards of the state, as young boys and old fools disturbed the birds so much that their number began to decrease.

From their colonies the blue herons leave in the morning for their feeding grounds. They often fly as far as fifty miles and more, but always return towards evening.

When it gets so cold that frogs, snakes, and slugs seek their winter quarters, and the fish retreat into deeper waters, then the cranes leave us, because they can no longer find their accustomed food.

§ 20. Larger Blue Flag. *Iris versicolor.*

MATERIAL: Blue flags with rootstocks, flowers, and seed pods. Observe: How insects fertilize the blue flags.

D

In April and early May the Marsh Marigold is the most conspicuous flower in cold, wet places; but in May and June every pond, every brook, is fringed by the well-known Blue Flags. Sometimes they grow in the water, sometimes near it; but always in wet places.

The stems and leaves grow from a stout, branched root-stock, which remains alive in the ground. In very wet and rich soil they grow about three feet high; while on soil which is rather poor and has insufficient moisture, they are often not more than a foot high.

The leaves are all sword-shaped, and near the ground, each lower one straddles the one above it. Where the leaves straddle, the side towards the stem is the upper and the one turned away from it is the lower side, and the leaf is simply folded upward along the sharp ridge. Where the leaves become free of one another, the two folds have grown together and the leaves now seem to stand vertically and not hori-zontally as on most other plants. A superficial observer would say they have a right and a left side, but only an upper and lower edge. The short leaves which grow near-est to the flowers are called bracts.

What is the most common number of flowers on a stem? The flower consists of three large outer divisions and three smaller inner ones. You must look close to find the pistil and the stamens. The pistil bears three leaf-like styles, and under these the stamens with their long anthers are hidden. The stigma or scar is quite concealed; but if you look sharply, you will find it on each one of the styles as a small membrane with delicate hair and covered with a sticky substance. You will see at once that pollen and stigma of the same flower cannot come in contact. But when a bumblebee wishes to get at the honey, it carries away some of the pollen of one flower and brushes it off on the stigma of another.

Find where the honey or nectar is in the blue flags. Patiently and closely observe the bees get it; lead the children to see and understand the process, and then describe it. Watch the ripening of the seeds and find out

FIG. 9. LARGER BLUE FLAG. *Iris versicolor.*

a, showing how a bumblebee enters the flower; *b,* flower with perianth removed; 1, stigmatic membrane; 2, a lobe of the style; 3, an anther. Both reduced.

how they are scattered. Do you know how seeds of the Dandelion and the wild Touch-me-not are disseminated? Compare other wild-growing flags, the Blue-eyed Grass and the Cultivated Iris. Let the children watch insects on Milkweeds. Catch butterflies on milkweed flowers and examine their claws closely. Small flies sometimes cannot get away from milkweed flowers. Try to find out what holds them.

§ 21. Review of Life in and near the Water.

If on your rambles to ponds, lakes, and meadows you did not walk with closed eyes, you must have been struck by the abundance of life, which seemed to be everywhere.

Poplars and Willows had taken possession of the always moist soil. A thick matting of grasses covered the low meadows; only here and there some flowers had managed to hold their own against the countless hosts of grasses. Cat-tails and many other rushes and reeds formed a perfect thicket in the shallow water. Bobolinks and blackbirds made meadow and rushes lively with their song.

The bottom of the lake was in some places entirely covered with small weeds, that never rise to the surface. Where the water is a little deeper, beautiful lilies floated their grand leaves and flowers in the sun.

And what a host of swimming, creeping, crawling, wriggling animals we find in this water. On the surface, under the leaves, among the weeds, and on the bottom they hunt and play.

On the very minute animals and plants the different kinds of larvæ, tadpoles, and small fish live, while they themselves hide under the leaves and in the weeds when pursued by their enemies, the large fishes and the birds.

All the plants mentioned above need a great deal of moisture. Very many animals are attracted to the water, because they also need it to live in and because the plants offer them food and shelter. If a lake or pond dries up permanently, all these plants die; the animals also die, if they cannot walk or fly to find another home.

The larvæ of mosquitoes and flies are eaten by fish. Flies

and mosquitoes are preyed upon by frogs and fish; and frogs and fish are often caught by kingfishers and herons. *Smaller animals often serve as food for the larger ones.*

Flies, mosquitoes, and frogs cannot be easily found in the winter. The first two hibernate in sheltered places, such as cellars and caves; the last pass the winter in the mud of lakes and ponds. As soon as the chilly nights make winged insects scarce the swallows leave us. When the frogs take up their winter quarters and ice begins to form on the water, the herons and ducks also move south. When do the different migratory birds leave your locality?

Plant life also sleeps through the long, cold winter in buds, roots, and seeds. By the warmth of spring plants are awakened to new growth and animals to renewed activity. Then the migratory birds return to their northern homes. When do they arrive in your locality?

III

GEOLOGICAL ACTION OF WATER

Visit a newly formed gulch, ditch, or washout. Field lesson; to be given a day or two after a heavy rain. See Farmers' Bulletin, No. 20, Washed Soils: How to Prevent and Reclaim Them.

§ **22.** When we visited this place a few weeks ago, this deep ditch was not here. Can you tell me who made it? You say the water did it; but I wonder where the soil is which was taken out here. You see so much has been taken out, that a dozen teams would have to haul dirt for a day, at least, to fill up this hole. Let us walk down this gulch and perhaps we shall find some of the washed-out material stranded somewhere. Here we are at the foot of the hill; just notice all the fresh stones, gravel, and sand here. Where did all this matter come from? Out of that washout, of course. Up on the hill we noticed how the water had carried material away; here we see how it has sorted this material. How could it do that? One day when we visited the brook you saw that the water rolled along small stones where the current was rapid. Now, when the torrent rushed down this run, it carried small stones, sand, and fine mud, all mixed together. As soon as the foot of this hill was reached, the velocity of the current decreased; it could no longer carry the larger stones and here they were dropped. A little farther down, we find smaller stones and gravel, and still a little farther down, you see the sand spread out. But what became of the fine

---

Observations. — Note the various grasses in meadows and in other places.

38

black soil, which has also disappeared from Mr. Jones' cornfield? We do not see it here, so we must go down to the creek; perhaps we shall find it there.

Indeed, how muddy the water is; it seems that a great deal of fine, corn-land soil has been stirred into this water. We will take a bottle of this water and stand it in a quiet place. But do you think this fine, dark soil in the creek comes from Mr. Jones' field? No, I think not; it came from places farther up the creek. And where does it go?

Look sharply about you; perhaps you can find the answer then. John thinks he knows. Let us see what he has found. In this small low place the muddy water stood for some time, and there is some yet in the deepest place. See how the soil is covered with a layer of fine mud. The water could carry this along although its current was quite slow, but when it became very slow or stood still, even this fine material was dropped.

We saw that the creek was quite muddy after the last heavy rain, and just as muddy we should find the river into which the creek empties. If you ever see some of our largest rivers, you will find their water quite muddy.

Streams always carry most material at the time of a freshet or flood. Wherever they overflow a swamp, a meadow, or dry land, their current slows up as the water spreads out, and there some of the material is deposited. The finer the material the farther it is carried, and the very finest is not dropped before the river enters the ocean. Here many large rivers, including the Mississippi, form large deltas with the sediment they carry. The delta of the Mississippi contains about fourteen thousand square miles. You can easily see that all deltas are grow-

---

Let the children see who can find the most species, and who can make the prettiest bouquets of the ears or heads.

ing; that of the Mississippi advances a mile in sixteen
years.

§ 23. The Sand Bank.

Visit a sand bank formed by a river or creek. If that is not accessible, the sand spread out below a washout will illustrate the lesson.
Have a gardener's trowel or spade to dig into the sand.

Field lesson. — From observations which we made some
time ago, you can tell me how this sand bank was formed.
Of course, the river brought the material from farther up
the stream and deposited it here. If I dig a small hole
into the sand, you will see that it is arranged in definite
layers or strata. As water spreads the fine sand or mud
quite uniformly, it could hardly be otherwise. Now we
will go back and dig a hole on the high bank, where the
water reaches only at the time of high freshets. Here,
also, we find the soil quite distinctly stratified. After high
water I have seen this bank covered with a layer of mud.
As the river has been here thousands of years, this must
have occurred many times, and in that way this bottom
land has been formed. Rivers and creeks, however, are not
only depositing sediment; they are also cutting away their
banks all the time. Often they deposit on the left side
when they cut on the right, and *vice versa.* By and by
some of the soil is washed into the ocean, as we have
learned. Most of that soil has been taken up and deposited
again and again before it reaches the river's mouth.

Now look at the rock which you see exposed on the bank
there. It is arranged in layers, which we can easily trace.
In the stone quarries which we visited we found the rock
arranged in layers. Wherever water runs over these rocks,
it cuts some of it away, turns it over and over, and grinds
it up into soil.

Observations. — **Try** to find the flowers of grain and of grasses.

But there is still another way by which rock is changed to soil. Water soaks into cracks and between the layers. When this water freezes it expands, and often breaks off large pieces of rock. Water also soaks into most rocks as it does into a sponge. Some of the rock is dissolved like salt and carried away, and by and by the hard rock crumbles into soil. You can see this decay of rocks wherever rock is exposed to the atmosphere and to rain. *By this slow decay and wearing down of rocks soil is made.* A large part of the soil in the Mississippi basin has been made in this way from rock in the Alleghanies and the Rocky Mountains.

Have you seen buildings or bridges where the rock is wearing away?

You see that rock sticks out on both sides of the river, and it extends for many, many miles under the loose soil into the country. In some places it comes to the surface, sometimes we strike it not more than ten to twenty feet below, at other places it lies at a depth of one or several hundred feet; but we should find it everywhere if we could go down deep enough.

We have just learned, then, that nearly all of the interior of the earth is composed of solid, hard rock. The loose soil compared with the whole mass of the earth is like a thin layer of dust covering a large globe. From observations made on high mountains and in deep mines, people have learned something about the rocks for a depth of about ten miles.

You have learned now how soil is formed by water, and how ice helps to split up the rock. Some day I must tell you what an important part large ice fields, called glaciers, have played in the formation of soil in our Northern States.

Do all grasses furnish equally good hay and pasture?

§ 24. How the Rocks were made.

To-day I will try to answer your questions about "How the rocks came to be." They were formed or deposited in water just like the bottom lands and sand banks we visited. You noticed that the soil grew more compact as we dug down, because the soil above, so to speak, pressed it. Deep, heavy waters pressed on the rock above, and the upper layers also pressed on the lower. This pressure was one of the forces which made rock of the fine mud or sand. In many rocks there is also some cementing material which holds the particles together, as lime cement binds the sand grains in mortar; and some rocks have been burned hard like brick. Most stratified rocks were formed on the ocean bottom, by mud or sand washed into them by rivers and waves. How do we know that? Much of this stratified rock extends in almost horizontal layers over thousands of square miles, and in it are often found fossils of marine animals. In some places stratified rock has been formed in large fresh-water lakes. *Whenever you find distinctly stratified rock extending over large areas, you may take for granted that it was deposited by water as mud or sand.*

I know what several of you are going to ask me now. You want to know where all this mud and sand could come from. Some day I shall tell you about that. You may find something about it in your geographies or in some simple treatise on geology.

If you can visit different kinds of stratified rock, study it closely. Can you easily break it with hammer and chisel, or is it so hard that it will scratch glass? Does it seem made of grains of sand, or does it appear like hardened clay? If you find many fossils in it, you most likely have a limestone. This rock has the peculiar "limey" taste and

---

Observations — Find the seeds of grasses.

smells like a freshly whitewashed room. Foremen and owners of quarries can probably tell you about the rock in your locality. About the bowlders which we find scattered all over our Northern States we shall learn later. Do you think that they were made here, or that they came from some other place? How could they be transported?

NOTE. — Observations made in the field and conclusions reached must be thoroughly reviewed in class.

IV

IN THE MEADOW. JULY AND AUGUST

MATERIAL : Different kinds of grasses from high and low meadows, from swamps, field, and garden. In presence of the class, wash the soil from a piece of sod. The names of species are not important in this lesson.

§ 25. When I asked you a few days ago how many different kinds of grasses you had found in the neighboring meadows, some of you were surprised to hear me speak about many kinds of grasses; but I see that you found more kinds than I had expected.

The most striking characteristic of the grasses is their growing close together in large numbers and forming meadows and lawns.

Most grasses have a hollow stalk, which is very high compared with its thickness. This stalk is called a culm. The long culm, with the ear or head, would scarcely be able to withstand the force of the wind, if it were not strengthened by some special devices. Let us take a sharp knife and make a number of cross-sections of the stem from the root upwards. We find that the lowest part of the stem is the hardest and the most compact. If this were not so, the culms would easily break or bend near the ground, because the force of the wind is, so to speak, concentrated there. Take a stick which is of uniform thickness and strength from end to end, fasten one end of it; then take hold of the other end and bend the stick until it breaks. It will break

Observations. — Plants of the cultivated sunflower; sow some sunflower seed and compare the seedlings with older plants.

44

near the fastened end. Why must the trunk of a tree be strongest near the ground? Why is a crowbar strongest near the lower end? The culms are strengthened still more by several other devices, two of which we can easily understand. You notice that the culms have knots, or nodes, at intervals, that these nodes are solid and hard, and that they are closest together near the ground. The leaves all surround the culm like a sheath for a part of their length. These sheaths strengthen the culm in the weak points just above the nodes. Strip off the leaf sheaths and prove that there are weak points just above the nodes. *The hollow culms thus strengthened admirably combine strength with lightness and, as we all know, are not easily kneed or broken before the seeds are ripe.* What change takes place in the strength of the culms after the seeds are ripe? Does ripe grain show boldly upright ears and culms?

Grass culms do not branch; their leaves are almost or entirely linear in shape and have parallel veins. Show how grasses are enabled to grow so close together. How many stalks can you count in a square yard of meadow?

Perhaps it has never occurred to you that grasses have flowers; but what we call ears or heads are really their flowers. None of them, it is true, have a distinct calyx and corolla; still, if we look just at the right time, we can easily find the tiny stamens with anthers hanging out of the little chaff-like scales. The pistils often end in delicate little feathers. You must try to find both. The heads of grasses make very pretty bouquets, which last a long time, and need not be put in water. We shall see how many pretty bouquets we can collect.

We have spoken several times about insects fertilizing flowers which attract them by their bright colors and by

---

Look for plants of the goldenrod, wild sunflower, rosin-weed, blazing star, wild aster, garden aster, dahlia, dandelion, and other composites.

their fragrance.　Now we find that the flowers of grasses
have neither bright colors nor fragrance, and that they are

FIG. 10.　BOUQUET OF GRASSES.

a, Redtop, *Agrostis vulgaris ; b*, Timothy, *Phleum pratense ; c*, Kentucky
Blue Grass, *Poa pratensis ; d*, Squirreltail Grass, *Hordeum jubatum.*
All reduced.

very seldom visited by insects.　Most of them are fertilized
by the wind.　When the tiny anthers dangle in the breeze,

Observations. — Examine the roots of the plants just named.

the pollen from one flower is taken up and wafted to the pistil of another. If, however, the wind is too strong, the whole anther may be shaken off before the little dust pouch has opened; in that case only a few flowers are fertilized and the yield of seed is small. Try to find a number of grasses in bloom and convince yourself that insects seldom, if ever, visit them. Can you name some cultivated plants which belong to the Grass family?

§ 26. Economy of Grasses in Nature and their Usefulness to Man.

Of all plants, the grasses are by far the most useful to man in northern latitudes. Our cereals — wheat, oats, barley, rye, and Indian corn — are simply cultivated grasses. Prove by the structure of these plants that they are grasses. The first three have been cultivated in the Old World from time immemorial, so that we do not even know now from which wild plants they were derived. Maize, or Indian corn, was cultivated by the natives of America when the country was discovered. Our well-known sorghum, the sugar cane of the South, and cultivated rice are also grasses.

If we examine the plants in a pasture, we find little else but grasses. What is the explanation for this? We must dig up some of the sod and wash the soil out of it. We find that besides the many fine roots, most grasses have a thick rootstock, which survives through severe winters and through parching droughts, and from this rootstock the blades and stalks grow in spring. Cattle and sheep often graze the pasture so closely that only very few grass stalks have a chance to produce seeds. *If all grasses had to grow from the seed every spring, our domestic animals would soon exterminate the very kinds which they like best.*

Do these plants grow annually from a subterranean rootstock or from seed?

Why are Goldenrods, Shoestrings, and even some grasses, allowed to flower and seed in pastures? Are our cultivated grasses annuals or perennials? What are Pigeon grass, and Millet?

The fact that most grasses are perennials, and that they grow so close together in large numbers, fits them peculiarly well to be the forage plants for millions of cattle, horses, and sheep. Our noble, wild, herbivorous animals, the deer and elk, feed largely on grasses; while the immense herds of buffaloes which not long ago roamed over the Western plains lived almost exclusively on grasses.

Our bread comes directly from the seeds of grasses; but indirectly even our butter, milk, cheese, meat, clothes, and shoes are derived from the grasses; for grasses are the principal food of the animals from which we get milk, meat, wool, and skins.

Millions of bushels of barley, rice, and corn are used every year in brewing beer, which is considered by many people an innocent and pleasant beverage, like coffee or tea; but the intemperate use of it cannot be too strongly condemned. Many million dollars' worth of rye and corn are used by distilleries in the manufacture of alcohol. Much of this is used in different arts, by scientists, and in medicine; but a large part of it is made into whiskey and brandy. These alcoholic liquors, taken habitually, are no doubt positively injurious, and they represent an enormous amount of grain, money, and labor thrown away.

Grasses in the economy of nature. — In the temperate zones grasses are the plants which give character to the landscape. They cover the prairies, meadows, and pastures like endless green carpets. By their many roots, they hold the soil together and prevent it from being washed or blown away. Besides feeding the animals already mentioned, they

Observations. — Observe weeds along roads, in fields, and gardens.

furnish food and shelter for countless numbers of insects, birds, and other small animals. *A country without grasses would be a desert.*

It is to be regretted that many of our farmers do not derive more profit from their meadows. On the continent of Europe high meadows are regularly fertilized with short or with liquid manure, and most low meadows are so laid out that they are thoroughly drained and that they can be flooded at any time. This flooding is done mostly on account of the fertilizing substances which the water carries with it.

NOTE. — About grasses especially valuable in your state or vicinity, consult the bulletins of your State Experiment Station. On Grasses as Sand-binders for Lake and Ocean Shores, see Yearbook of the Department of Agriculture, 1894.

E

V

PRAIRIE FLOWERS IN AUGUST

MATERIAL: Several species of goldenrods, blazing star, wild sun-flowers, cultivated sunflowers, prairie weeds; several kinds of aquatic plants for comparison. If no wild prairie is accessible, a piece of high ground not covered by trees or rank road-weeds will do for the intro-ductory field lesson. Strips of wild prairie can often be found along railroad tracks.

§ **27.** Some weeks ago we spent considerable time in studying plants and animals found in wet places or in the water. To-day we shall take up a group of plants which make their homes in open fields and on dry prairies. Here in my left hand I hold the flowers we picked yesterday along the railroad track, and in my right hand I hold some of our friends from the lake bottom. You all have similar specimens on your desks. Notice the difference in the general texture. The aquatic plants are soft, easily cut and compressed. Try to make them stand upright. In their lake home they grew erect, because the water buoyed them up, but in the air they fall over and shrivel up.

If you make cross and longitudinal sections of stems and leaves of aquatic plants, you can see quite well that their whole tissue is composed of cells. All vegetable tissue is composed of similar cells, but in most other plants the cells are much smaller, are closely packed, and often have com-paratively thick walls, which make the plant hard and rigid.

Observations. — What weeds do you find growing along roads, in fields, on vacant lots, and in other waste places?

Now notice how strong and robust our prairie flowers are. The Goldenrods, the Blazing Star, the Sunflowers,

FIG. 11. BOUQUET OF PRAIRIE FLOWERS.
a, Purple Coneflower, *Echinacea purpurea*; b, A Goldenrod; c, Blazing Star, *Liatris scariosa*. All much reduced.

the Asters, and others stand upright without any difficulty. Their stems are almost woody, their leaves are more or less rigid, and are often covered with fine hairs.

What are eight of the more common weeds of your vicinity?

Prairie plants need a strong stem to withstand the force of the wind. Aquatic plants generally grow in sheltered bays or in deep water, where winds and waves affect them but little. Those that grow in running water float and are wafted about by the current, and simply cling to the bottom by their roots. *Aquatic plants in their peculiar environment do not need a strong, rigid stem; it would even be a disadvantage to them.* Can you show how?

Here is a handful of aquatic plants which I exposed to the sun and air for a few hours; and here you see a handful of prairie plants, which I exposed in the same way. Notice how the water plants have shrunk and shrivelled, while the others are not much the worse for the exposure.

You know that prairie plants are often exposed to drought and dry winds, and therefore have learned to endure considerable drying. Aquatic plants evaporate their moisture very rapidly when taken out of the water, and shrivel so much that little is left of them.

Goldenrods, Blazing Star, Coneflower, and many others, which in form remind us of Asters and Sunflowers, are characteristic Prairie Flowers. When these beautiful children of our boundless plains mingle their gold, silver, and purple hues with the verdure of tall grasses, we are gently reminded that soon nature will once more be sleeping, while the "Northwester" sweeps furiously over her white blanket.

Examine one of your flowers carefully, and you will find that each head is composed of many very small flowers, which grow from a common floor. This composite flower is surrounded by a common calyx, which often consists of many green scales or bracts. Nearly all of these plants are hardy, and a light frost does not kill them.

Plants which have such composite flowers are called

Observations. — Are the following among them?

Composites. They form the largest family of flowering plants on earth. Do you think there are many of them growing in this locality? For to-morrow's lesson, bring me as many different kinds of composites as you can find. Do all Asters, Thistles, and Dahlias belong to the composites? How are the seeds of Thistles and Dandelions disseminated?

In our next lesson we shall study the cultivated Sunflower; it is the largest of our composites and shows the typical structure very clearly.

§ 28. The Sunflower. *Helianthus annuus.*

MATERIAL: Cultivated sunflowers at different stages of development; other composites for comparison.

The cultivated Sunflower might well be called the Giant Flower. In good soil you can find some whose flower heads measure from six to ten inches across. If the plant has room, it is copiously branched; its stem is then about as thick as a man's arm, and may attain a height of seven feet.

Each head consists of a large number of little tubes and a row of strap-shaped rays on the outside. A single one of these little tube-shaped flowers, if placed by itself, would not be at all conspicuous, and could not easily be seen by insects. *The fact, however, that thousands of them are placed in one head, and there surrounded by a crown of large, bright rays, makes the whole head very conspicuous.* On a warm day you will find a large number of bees and other insects busy drinking the honey from the thousands of little wells.

The little tubes open in rings, beginning on the outside of the head. Look at this large head. Here is a ring of

---

Common ragweed, giant ragweed, false sunflower, dog fennel, Canada thistle, burdock, sweet clover, Russian thistle.

tubes from which protrude little bodies, looking like horns.
Inside of these is a ring, in which every tube bears a hat
of yellow pollen. If, now, a bee crawls about on the sunflower, it gets its head all dusty, when it introduces its long
lips into the tubes. With its body it brushes away the
little pollen caps, and scatters the dust all over the curved,
horn-like bodies. These are the pistils, and *in this way* the
insect carries the pollen of one tube to the pistil of another.

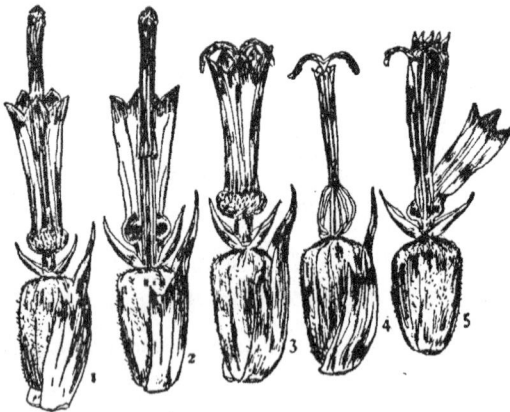

FIG. 12. FLORETS OF CULTIVATED SUNFLOWER.

1, showing the projected anther tube with pollen on top; 2, the same,
with half of the corolla tube removed; 3, showing the two tubes and the
style; 4, the same with the corolla tube removed, and showing how the
anther tube is retracted; 5, half of the anther tube and half of the corolla
tube removed. All enlarged.

Insects, no doubt, also carry pollen from one head to another,
and also from one plant to another.

The parts of the florets. — With a pin or small knife, open
one of the tubes lengthwise. *The tube itself is the corolla.*
Inside of it you will find little threads grown together at
their upper end and forming a second short tube. *The
threads are the stamens, and the second tube is 'formed by the
growing together of the anthers.* In the centre of the stamens
is the style. It grows through the anther tube and pushes

the pollen out before itself. Thus the yellow hats are formed.

But the style is not yet split, and is not ready to be fertilized. Only after most of the pollen has been brushed away by insects does it split, and the upper sides of the little horns, which are yellow or dark purple, are now ready to be fertilized by pollen from the florets.

Students of botany have found that it is an advantage to most plants if their flowers do not fertilize themselves, but are cross-fertilized by insects, or some other agency. *In cross-fertilization the pollen of one flower is carried to the pistil of another. Many flowers have exceedingly fine and interesting devices for preventing self-fertilization and for securing cross-fertilization by insects.* In some plants, as, for instance, in our well-known milkweeds, self-fertilization is absolutely impossible. Some flowers are self-fertilized only when insects are scarce; others are wind-fertilized; and some are always self-fertilized. Can you tell now by what means the sunflower secures insect-fertilization and prevents self-fertilization? Do the pistil and the anthers in one floret mature at the same time?

On top of the immature seeds you discover two small scales; these really are the calyx for one floret. In many composites this calyx is changed to hairs, or feathers, or awns; it is called pappus.

The common floor on which the florets are set is called the receptacle. It is often covered with chaff-like scales, which probably protect the seeds and help to hold them in place. We have already learned that all the florets are surrounded by a common calyx, which consists of rather large leafy scales, often arranged like shingles on a roof. This common calyx is called the involucre.

Have a druggist weigh for you one sunflower seed. You will find that it weighs about a gram. Then weigh on a common spring-scale a large, full-grown sunflower plant, with roots, leaves and all. How many times is the weight of the seed contained in the weight of the mature plant? In how long a time does this wonderful growth take place? If your plant is three months old, how much was its average gain in weight per day? Where did this rapidly increasing material come from? Will a plant grow if you cut off all its leaves? Make the experiment.

§ 29. The Composite Family.

MATERIAL: Sunflowers, goldenrods, asters, blazing star, and other composites of the season brought by the children.

Here we have many flowers of the season, some very showy and a few quite modest in appearance. In my left hand I hold ten flowers, differing much in size and color; but all agree in one thing. *Their larger heads consist of many tube or strap-shaped florets, growing from a common receptacle and surrounded by a common involucre. They all belong to the Composite family, of which we studied the Sunflower as the type.*

Here are some thistles and blazing stars. Examine their florets and you will find them all tubular. Here is a dandelion, which has all the florets strap-shaped. Other flowers resemble the sunflower very closely. They have a margin of large, showy rays; but the inner florets are tubular. Can you tell to which one of these three divisions each one of our composites belongs?

Müller, in his famous book on "The Fertilization of Flowers," mentions the following points, which combined so well to fit the composites to survive in the struggle for existence: —

1. The close association of many florets. This makes them more conspicuous and attracts more insects. Insects can fertilize numerous florets in a very short time. The

common involucre makes a separate calyx for each floret unnecessary and it is generally changed into feathers, hairs, or barbs, which aid in the dispersion of the seeds.

2. Their honey is easily accessible and attracts a great variety of insects. Look for a ring around the style of the sunflower, at the base of the tube. This ring secretes the honey. Have you found the honey in the sunflower? If not, try to find it. How is it protected from rain? Find the honey in milkweeds.

3. By an arrangement described in the last paragraph, cross-fertilization is very generally secured. Describe that device again.

The Composites are a characteristic feature of the landscape in all of eastern North America. You can find them from spring until late in the fall, but in late summer and in autumn they are most conspicuous. And now we must say good by to them all, — to the Garden Asters and the Dahlias, which make our lawns shine in all the colors of the rainbow, and to the Goldenrods and native Asters, the wild and hardy children of our wild hills and prairies.

VI

ROADSIDES AND NEGLECTED CORNERS

§ 30. A Chapter on Weeds. How they grow and how you can kill them.

NOTE. — Of the following list choose for study only those which are common in your neighborhood or which are likely to invade it.

1. Common Ragweed, *Ambrosia artemisiæfolia.*
2. Giant Ragweed, *Ambrosia trifida.*
3. False Sunflower, *Iva xanthiifolia.*
4. Dog Fennel, *Anthemis coluta.*
5. Canada Thistle, *Cnicus arvensis.*
6. Burdock, *Arctium Lappa.*
7. Sweet Clover, *Melilotus alba.*
8. Russian Thistle, *Salsola Kali.*

These weeds can be studied at any time from July to October. Teach the children to recognize them at all stages of development, beginning with the seedling. The following descriptions refer to the plants in flower.

Nos. 1 to 6 belong to the Composite family, but the flowers in the first three are small and greenish, although very many of them are set close together. The two Ragweeds have the staminate and the pistillate flowers in separate heads. Both are exceedingly common in all our Northern States east of the Rocky Mountains. It would be difficult to find a road in this territory along which these two weeds are not found in profusion, if they are left undisturbed.

The Giant Ragweed grows from one to twelve feet high; the Common Ragweed attains a height of about three feet. The

Observations. — The great abundance of house flies, moths, and all kinds of insects.

latter has the leaves much cut up into narrow lobes, which resemble small rags.   The former has very large leaves which are deeply three-lobed.   Can you tell now why these plants are called Ragweeds ?   Both of them are annuals. How can you prove that ?   Their seeds are dispersed by the wind and sometimes with grain.   Both do some injury to grain, if they are ·allowed to grow in it.

Fig. 13. Leaf of Giant Ragweed.
Much reduced.

Fig. 14. Leaf of Common Ragweed.
Much reduced.

No. 3. *The False Sunflower.*

This is another very large weed.   It is found in cultivated soil, along roads, and in waste places, where it grows from two to six feet high and the stem may become from one half to two inches thick.   The leaves resemble closely those of the sunflower, but their margins are more cut-toothed.   Do you think these three weeds are fertilized by insects ?

The various kinds of spiders and how they secure their prey.

How to exterminate them.

As all three are annuals, they are best killed out by preventing them from seeding.   Cutting or pulling will accomplish the result.   If the plants have already matured seed, they must be burned, after cutting.

No. 4. *The Dog Fennel.*

This weed also is a composite and at once recognized as one. Its flowers have white rays and a yellow disc (the central part); the leaves are very finely divided.   The whole plant has a strong and rather unpleasant scent.   It is often a troublesome weed in meadows and pastures.   Early mowing, before the plants have seeds, will exterminate it.   Why do cattle not keep this weed down? How are the seeds of this species disseminated?

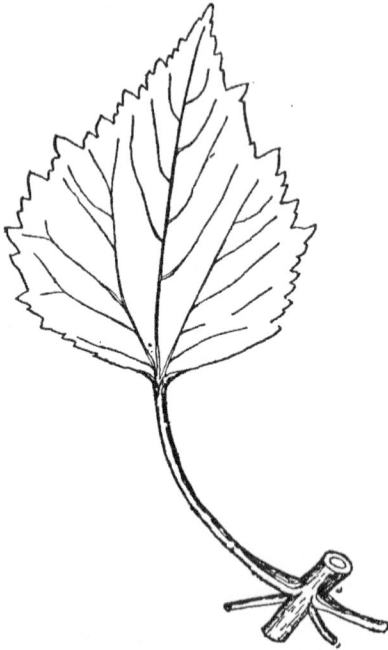

FIG. 15. LEAF OF FALSE
SUNFLOWER.
Much reduced.

No. 5. *The Canada Thistle.*

This is one of the worst weeds in our Eastern and Northern States, and in Canada.   It was introduced into Canada from Europe, and from Canada it has spread into the United States.   It is now found from New England to Missouri, and also on the Pacific coast.

The plant is one to two feet high, has rose-purple flowers, which are smaller than those of the Common, or Bull Thistle; from which it also differs in its more slender

Observations. — Bats flying about in the evening catching nocturnal insects.

habit, thinner, more deeply cut and curled leaves, and less rigid prickles. The teacher should show both plants at the same time.

FIG. 16. LEAF FLOWERING HEAD, AND ROOTSTOCK OF CANADA THISTLE.
One-half natural size.

Note the fall migration of birds ; hibernation of amphibians, reptiles, insects.

The Canada thistle is a perennial, which means that its roots live in the ground from year to year. In spring these roots send out shoots, or rootstocks, in all directions. A single plant can spread over a square rod in two or three years by its rootstocks alone. This is the reason that the weed is so difficult to eradicate. Every owner of land should see that it does not become firmly established. Frequent grubbing or mowing, ploughing several times in August, salting the plants and then pasturing sheep on them, applications of kerosene and carbolic acid, are all recommended as means to kill the Canada Thistle. About the detail of those measures consult your State Reports or Bulletin No. 39 of the Wisconsin Agricultural Experiment Station.

No. 6. *The Burdock.*

The Burdock is another common weed, too well known to need any description here. In size it rivals the giant ragweed and the false sunflower. It is not troublesome on cultivated land, but takes possession of every corner of rich soil, where a lazy and careless owner leaves it undisturbed. What makes it a nuisance? Can you tell to which family it belongs?

This weed is a biennial, which means that it lives two seasons or summers.

The seeds germinate generally the spring after they matured. During the first summer the plant simply produces a whirl of large leaves and a strong root, but no stem. This root lives through the winter, and in the following spring, the second in the life of the plant, produces leaves, a branched stem, and a large number of flower heads. When the seeds have matured, these heads form

Observations. — What becomes of house flies, wasps, and mosquitoes in late autumn?

FIG. 17. LEAVES, FLOWERING HEAD, AND HEADS WITH SEED OF
BURDOCK.

One-half natural size.

Do all birds leave us in autumn?

the burs, which are so annoying to man and beast. Can you tell now how the seeds of the burdock are disseminated ?

How destroyed. — The burdock dies a natural death at the end of the second season. Observe such dying plants. During the first season, the root can be cut under ground or the whole plant may be pulled up, when the ground is very wet. Try that. Plants that have lived into the second summer should be cut repeatedly, to prevent them from blooming and producing seed. The burdock, like the Canada thistle and the dog fennel, was introduced from Europe.

No. 7. *The Sweet Clover.*

This plant is not troublesome in fields and gardens. In Minnesota and Wisconsin it grows luxuriantly along roads, on almost barren limestone, and in other waste places. In its general appearance it resembles alfalfa, but is much larger, being from two to seven feet high. It can scarcely be classed as a noxious weed; its numerous white flowers are quite fragrant and the plant has been cultivated for fodder. If, however, it is allowed to grow along roadsides, its profuse stalks and branches offer good shelter for hibernating insects. Repeated mowing will keep it from seeding, and ploughing or tilling of any kind will soon destroy it. The flowers of the sweet clover resemble those of white and red clover. If you can compare the different clover flowers with flowers of beans and peas, you will notice a close resemblance. Beans, peas, vetches, and clovers belong to the Pulse family. Do you know a tree whose flowers resemble the flowers of the pea and the bean ?

In some places a Yellow Sweet Clover is found, which in general appearance is much like the white species, but is smaller, from two to three feet high, and has yellow flowers.

No. 8. *The Russian Thistle.*

This is one of the most injurious of the weeds that have been introduced into the United States. In the following description I quote from Mr. E. S. Goff in Wisconsin Bulletin, No. 39· "The Russian Thistle is an annual, coming each year from the seed. It grows from a single, small, light-colored root less than half an inch in diameter and six to twelve inches long to a height of from six inches to three feet, branching profusely, and when not crowded, forms a dense, bush-like plant from two to six feet in diameter and one-half or two-thirds as high. When young, it is a very innocent-looking plant, tender and juicy throughout, with small, narrow, downy, green leaves. When the dry weather comes in August, this innocent disguise disappears, the tender downy leaves wither and fall, and the plant increases rapidly in size, sending out hard, stiff branches. Instead of leaves, these branches bear at intervals of half an inch or less three sharp spines, which harden but do not grow dull as the plant increases in age and ugliness. The spines are one-fourth to one-half inch long. At the base of each cluster of spines is a papery flower about one-eighth of an inch in diameter. If this be taken out and carefully pulled to pieces, a small, pulpy, green body coiled up, and appearing like a minute green snailshell, will be found. This is the seed. As it ripens, it becomes hard and of a rather dull gray color. At the earliest frost the plants change in color from dark green to crimson, or almost magenta, especially on the more exposed parts. When the ground becomes frozen and the November winds blow across the prairie, the small root is broken or loosened and pulled out. The dense, yet light, growth and the circular or hemispherical form fit it most perfectly to be carried by the wind. It goes rolling across the country at racing speed, scattering seed at every bound, and stopping only when the

F

FIG. 18. RUSSIAN THISTLE.

a, young shoot; *b*, mature spray; *c*, seed. *a* and *b*, slightly reduced;
c, enlarged.

wind goes down, or when torn to pieces; for there are few fences or forests to stop its course in the Dakotas."

Injury caused by the Russian thistle. — Where allowed to grow, it takes complete possession of the land and chokes grain and flax. It is injurious to corn, potatoes, and all cultivated plants. The rigid bushy plants, closely beset with sharp spines, are very disagreeable to handle.

Introduction and distribution. — " It was introduced from Russia in 1873. During the twenty-two years since, it has spread with greater rapidity than any other weed. Ninety new localities were reported to the Department of Agriculture during 1894, and sixty-nine in 1895, previous to November 20. It now occurs in places from eastern Ontario and New Jersey to the western border of Idaho and California, and from Manitoba to New Mexico and Missouri, being most abundant in South Dakota and adjacent states. Twenty states and three Canadian provinces are known to be infested." [1]

Remedies. — As the plant is an annual, it should be killed out before it has produced seeds. For a detailed account of this weed, see your State Reports, or " The Russian Thistle," by L. H. Dewey, Department of Agriculture, or Wisconsin Bulletin, No. 39. Is the plant really a thistle?

It is very commonly found along railroad tracks, about elevators and railroad yards. How do you account for its presence there? Do you think the seeds were carried to the Eastern States by wind? How could railroads distribute weeds?

See also on weeds : Farmers' Bulletin, No. 28 ; Weeds, and How to Kill Them ; Two Hundred Weeds, by L. H. Dewey, United States Department of Agriculture ; Bulletins Nos. 50 and 57 of the Kansas State Agricultural College (No. 50 contains one hundred and forty-five figures of weed seedlings, and No. 57 figures the leaves of about two hundred weeds).

---

[1] From L. H. Dewey, U. S. Department of Agriculture.

§ 31. Review and Summary.

MATERIAL : A piece of soil with its natural sod ; a piece of field soil with the stubble or grain plants.

Most of the wild plants growing in our country are natives of this continent.   The first settlers found them growing wild as we find them to-day.   Among cultivated plants, Indian corn and tobacco are also natives of North America, and were to some extent cultivated by the Indians.

Nearly all our other cultivated plants were brought from Europe when this continent was settled by white people. With the seeds of useful plants, nearly all the noxious European weeds have been introduced on our farms, and some of them, like the Russian thistle, have spread with amazing rapidity.   Do you know of injurious animals which have been introduced into this country ?

Cultivation has brought about very great changes in this country and especially in the surface of the soil.   Millions of acres, indeed whole states that formerly were dense forest or wild prairie, have been converted into farms and gardens. *The ploughed and manured land is good soil, not only for wheat and other grain, but also for a host of weeds.   These seed themselves, and man tills the ground for them unintentionally.*   In a wild country, such plants as the ragweeds and the false sunflower, which grow best on loose and bare soil, have to be content with the little ground which is laid bare or deposited by water runs, streams, and burrowing animals, such as gophers, skunks, rabbits, etc.; but when that wild country is settled, an unlimited acreage of ground is prepared for these weeds on farms, in gardens, and on roads and railways.   Of this favorable ground our native weeds quickly possess themselves, and their relations, introduced in various ways, were not slow to claim a share.

The fact that some farmers have more land than they can

thoroughly cultivate has done much to spread some of our worst weeds.

Still, in the economy of nature, weeds are not useless. They hold the bare and loose soil together, and keep it from being blown or washed away. Those kinds that grow on almost bare clay, sand, or rock no doubt add slowly to the fertility of such soil by their decaying roots, stalks, and leaves, thus tending to fit the soil for more useful plants.

VII

LIFE ABOUT OUR HOMES IN FALL.
SEPTEMBER AND OCTOBER

§ **32.** *Introduction.* — One who has seen a board shanty set down on the open prairie will at once admit that shade trees are absolutely necessary to make any place look like home. In the autumn our trees begin to show some yellow leaves, the merry songsters of spring have mostly left them. Noisy crowds of English sparrows frequent them in the day-time, and in the evening we may hear the low chirps of bats, which circle around them hunting for moths and other nocturnal insects. On bushes, and in corners, we find the webs and nets of different spiders which are waiting for flies to become entangled in their nets.

§ **33.** The Box Elder, or Ash-leaved Maple.

MATERIAL: Twigs with leaves and fruit, flowers from your herbarium, seedlings of last spring; piece of the wood cut slantingly, with the rough bark on it, observed in spring; the unfolding of the leaves and the appearance of the flowers.

The Box Elder is one of the trees most commonly planted for shade and ornament. It is a rather small tree, reaching an average height of about thirty feet. If you compare the fruit of this tree with those of the maples growing in your neighborhood, you will at once be struck by the close resemblance they show to each other; the Box Elder is, indeed, a kind of maple. The leaves consist of from three to

---

Observations. — Give special attention to your house plants.
70

five leaflets. The latter are attached to a strong midrib, which bears an odd-numbered leaflet at the end. A compound leaf, which has the leaflets attached along a common stalk,

FIG. 19. BOX ELDER.

a, part of leaf; *b*, seeds; *c* and *d*, staminate flowers; *e*, pistillate flowers. All reduced.

What plants do you find in the homes of your friends?

is called a pinnate leaf. Good examples of pinnate leaves are the leaves of roses, peas, and vetches.

The flowers of the Box Elder appear early in spring, with the leaves or before them. The sterile and the fertile flowers are on different trees. The sterile ones appear in drooping bundles; they have no corolla, the stamens are attached to a long thread and surrounded by a small green calyx. After the pollen is shed, the sterile flowers wither and drop off.

The fertile flowers appear in drooping clusters; they also are destitute of a corolla. Each flower has two styles which look like a small, bent horn. The fertile flowers develop into seeds, and when the seed is full grown, each has developed a wing. The seeds are not ripe until late in the fall. They drop off very slowly, and some are still on the tree the next spring.

The Box Elder, although its native home is the borders of streams and lakes, will thrive in almost any kind of soil, in this part of the country, although, like nearly all trees, it prefers moist and loose soil. It grows very fast. In one season young trees commonly produce shoots three feet long. Its foliage is very dense, and, if the young trees are not planted too far apart, they soon kill weeds and grasses by their dense shade. The fact that the tree remains rather small, makes it more desirable near buildings, because, in case of a storm, it does not endanger them by falling limbs. *These qualities make the Box Elder one of the most desirable shade trees in our Northern and Northwestern States.*

The wood of the Box Elder has no special value. Find out for yourself about its color and hardness.

NOTE. — The insects principally injurious to this tree are the Box Elder Bug, the Box Elder Leaf Roller, and leaf-lice. About these

Observations.— What are some of the plants most easily raised in houses?

insects, and remedies against them, consult : Report of the Minnesota Entomologist for 1895.

§ 34. The English Sparrow, or House Sparrow.

MATERIAL: One or two live specimens in a cage. Recently killed sparrows or good pictures may be substituted ; nest and eggs of sparrows. Previously observed: Sparrows molesting martins, bluebirds, wrens, bank swallows, robins, catbirds, and other beneficial native birds ; what the sparrow eats.

This bird is so common in nearly every town and village of the region for which this book is intended, that a description is necessary only for recently invaded localities.

It is about as large as a canary bird, but stouter. Its general color above is ashy, with black and chestnut stripes on the shoulders and back. Over the eyes and on the sides of the neck a dark chestnut mark can be distinctly seen. The wings are also marked by one chestnut and by one light bar bordered by a black line. The dark bill is cone-shaped and very strong. The feet are brown, the tail gray and there is a dark mark on the breast. The female is paler and does not show the bars and marks distinctly.

Its history in America. — Some poorly instructed would-be benefactors of their country imported eight pairs from England in 1850, and liberated them at Brooklyn, N. Y. It was thought that the sparrow would exterminate various insects injurious to shade trees in the streets and parks of cities. A regular sparrow craze seems to have seized the country in the last half of the sixties, as it is known that within that period sparrows were imported to New Haven, Conn., to Boston, Mass., and to Galveston, Tex. The city government of Philadelphia imported a thousand sparrows in one lot, in 1869. Within the next five years they were imported to San Francisco, Cal., Salt Lake City, Utah;

How are house plants generally propagated?

Halifax, Nova Scotia; and to several towns in Ohio, Wisconsin, and Michigan. It was not until about ten years later that the sparrow importers began to realize what a nuisance they had inflicted upon their country.

How the sparrows have spread. — As sparrows are attracted to grain-elevators, railroad freight houses and yards, it often happens that they are locked up in freight cars, where they frequently roost. When the cars are opened at their points of destination, the liberated sparrows are at once at home again, although they may have travelled hundreds of miles. They also fly from town to town, along railroad tracks, and when the smaller towns become crowded, they invade the country as well.

What they eat. — If you have observed them for some time, you must have noticed that they are not at all particular about their diet. In towns, they live chiefly on the undigested grain they find in horse-droppings, but they will also eat all kinds of kitchen scraps, buds of trees and shrubs, berries and fruits, grain on the field, different insects, except hairy caterpillars, and anything that is eatable. I have even observed them fishing. One day in June, 1895, as I was sauntering along the docks of the Milwaukee River, near its entrance into Lake Michigan, I noticed a considerable quantity of small pieces of wood and bark slowly drifting towards the lake. The water was very turbid and had a foul smell; and a large number of fishes from two to three inches long were floating dead on the surface. Wherever a dead fish came near a piece of wood or bark, the sparrows would alight on it and seize the fish; some eating their catch on shore, others apparently carrying it off to their young. Often a sparrow would alight on so small a raft that he had to keep his wings in motion, while he picked up the desired minnow. That

Observations. — Are they all grown for the sake of the flowers?

a bird with such perfectly omnivorous habits finds his table spread wherever man lives, is easily understood.

The house sparrow versus our native birds. — The courage and pugnacity of this little fellow are not at all inferior to his appetite and digestive powers. They will fight robins, wrens, catbirds, swallows, orioles, and any of our native singers. If the enemy is too large for one sparrow, a mob of them will molest him, until he leaves the neighborhood. They often drive wrens, martins, cliff and bank swallows, out of their nesting places, and appropriate the nests to their own use. A pair of robins and a pair of sparrows once built their nest on the same oak tree in my yard. At four in the morning, the birds began to chatter and fight, and continued to do so, with some interruptions, all day. At the end of a week, the robins left, and the sparrows stayed. Whenever sparrows become numerous in a town, our native birds become scarce.

As the house sparrow renders but very little benefit to us, and crowds out our beautiful and beneficial singers, it should be relentlessly destroyed. Break up their nests, shoot them, trap them, poison them, whenever and wherever you can. They may be poisoned in the following way: mix one part, by weight, of white arsenic with fifteen parts of wheat. Moisten the wheat, before stirring in the arsenic. A little gum arabic or mucilage added to the water will make the poison adhere better. Dry the wheat before using it. Attract the birds by feeding them regularly in a certain place; when they come regularly to the feeding ground, give them the poisoned grain. That the poisoning and shooting should not be entrusted to small boys, is self-evident.

You will find that it is by no means easy to shoot, trap, or poison sparrows, after their suspicion has once been

Visit a greenhouse, if you live in the city.

aroused. If you use Paris green or London purple instead of white arsenic, the birds will generally not touch the poisoned grain, on account of its suspicious color. Does this speak well for their intelligence?

Nesting places. — Martin houses, wren boxes, holes of bank swallows, ready corners, and holes about buildings and bridges are preferred; but when these cannot be had, the sparrow at once rises to the occasion, and builds on trees and in vines. I have found about a hundred nests under a wooden bridge in St. Paul, Minn. In this section of the country, they begin nest-building in March or April, and hatch several broods during the summer.

In Canada, New Zealand, and Australia, where the house sparrow has also been introduced, it has proved about as much of a nuisance as in the United States.

When we consider that this little bird is truly omnivorous; that its courage and pugnacity lead it to make excellent use of its stout bill and strong claws; that it can acclimatize itself in all temperate regions of the earth; that it can build its nest anywhere and out of anything; that it is a prolific breeder; that it is possessed of great intelligence to recognize danger and of no less cunning to avoid it, — we must perceive at once, *how exceptionally well it is equipped in the struggle for existence,* and why it has become almost cosmopolitan.

§ 35. The Bat.

MATERIAL: Any bat found in your region. The animal alive in a trap, or a picture of it. Previously observed: How bats fly about in the evening, how they rest, etc.

To the animals which are quite common and still but little known belong the different species of bats which inhabit the United States and Canada.

Observations. — Why will some greenhouse plants not do well in your home?

When the sun has set, the bats come out of their hiding-places and flutter about trees and houses, uttering from time to time a shrill chirp.

The general color of bats is that of mice. Some kinds are brown or reddish, others rather black, and their fur is as soft as that of a mouse. The little animal resembles a mouse so much in color, size, and general appearance that the Germans call it "Fluttermouse" (*Fledermaus*). But bats are very different from mice in several respects. *They have a peculiar kind of wings with which they can fly, and their teeth are quite different from those of mice.* We have seen how the legs of the frog are made for jumping and

FIG. 20. A BAT.

swimming. The limbs of the bat are so changed that they have become organs of flight.

A thin skin, or mantle, connects the fore and hind legs, and is also attached to the tail. The four long thin bones in which the fore limbs end are really the four digits, or fingers, much lengthened. Only the thumb of the fore limbs is short and bears a claw. Does the mantle remind you of a parachute or umbrella? With this parachute the bat can fly almost as well as the bird flies with his wings.

When bats fly about, they hunt for moths and other nocturnal insects, of which they eat a great many. Flying makes them very hungry, just as it does the birds.

The bat's eyes are small, but its hearing is very acute. *Its mouth opens almost up to the ears, which helps it to catch insects on the wing. The teeth are well adapted to insect food.* They are all sharp, and look like the points of so many needles. With such teeth the creature can easily hold and kill even the hardest beetle. What birds catch and eat their food on the wing? Do their bills open very far?

How the bat sleeps and rests. — The toes on the hind feet of a bat are free from the mantle. With these the bat hooks itself to a nail or twig, or to a rough wall, and thus rests with its head hanging downward. When evening comes, it lets go its hold, and while dropping spreads its parachute and flies off.

On the ground, bats are very awkward. — Their limbs are not intended for running, as are those of the nimble mouse. By means of the thumb-nails and the claws of the hind feet, they pull and push themselves along. They cannot rise from the ground, but climb up to some high place from which they drop down while spreading their parachute.

How bats live in winter. — When severe frosts have killed insect life or driven it into its hiding-places, the bats also disappear. Some migrate like birds, others hang themselves up in caves, rocks, crevices, hollow trees, and other places of shelter. The wings, by means of which the bat glided noiselessly about in the orchard on so many warm summer evenings, now become the cloak of the little creature, whose food supply has given out, and who has nothing better to do than to sleep away the long, dreary winter.

§ 36. Flies.

MATERIAL : House flies, stable flies, and blowflies in a bottle. Only a few of each species are needed.

Everybody knows the house fly. In late summer they are often so numerous that it is almost impossible to keep a house free from them.

Like all true flies, it has only two wings. Behind the wings, where other insects, as butterflies, bees, and wasps, have another pair of wings, the fly has two short stubs, which look as if a pair of wings had been cut off from them. Examine the flies mentioned above, to see if this is true of the stable fly, and the blowfly as well.

Life history of the house fly. — It lays its eggs in manure, and the larvæ look like the typical maggots that everybody has seen on dead animals, decaying meat, on cheese, and in old mushrooms. One fly lays more than a hundred eggs. The maggots become full grown in about a week. Then they cease eating and moving, their skin shrinks, and they look like little brown barrels. Whenever manure is hauled out of the barnyard in summer or fall, these little barrels can be seen. They are the pupæ of the flies. After the pupa stage has lasted also about a week, the maggot changes into a fly, which breaks the brown skin and crawls out. If you can procure some of the pupæ and keep them in a glass jar with moist soil, you can watch this change taking place.

Some flies winter in houses or warm barns, and when spring comes, they rapidly multiply.

The stable fly looks very much like the house fly, but it is grayer in color; its wings spread more, and it has a piercing beak like the mosquito, and annoys horses and cattle very much. When stormy and rainy weather approaches, it often comes into our houses and torments us with its biting. Most people think the house flies are hungry and do the biting. Examine the house fly closely, and you will find that its beak has a foot-like appendage, with which it can pick up small pieces of food and rasp on them, but cannot pierce the skin of men and animals. No house fly ever bit.

If flies were not so numerous, they would not annoy us; but on account of their immense numbers they often become a great nuisance, because they fall into everything, soil our houses and furniture, trouble us at table, and disturb us in other ways.   If barns and barnyards were kept clean, so that flies could not find so many breeding places, and if no refuse were thrown out near our homes, they would not be, nor become, so numerous.   Can flies see well?   Do you think they can smell?   How would you prove it?   What remedies do you know against flies?   Which would be the most thorough remedy?

The different kinds of blowflies are larger than the house fly, and are known by their loud buzzing.   They deposit their eggs on meat and dead animals.   If a fish is placed in the sun, the openings under the gills are filled with eggs in a few hours; so quick are the flies to find food for themselves and their maggots.

NOTE. — As the eggs and maggots of flies are revolting objects to most children, it is not advisable to show them in class.

§ 37. The Spider.

MATERIAL : An orb weaver, a cobweb weaver, and a funnel-web weaver, each by itself in a rather large bottle.   Place a few straws or small sticks in the bottles, and collect the spiders a few days before you use them.   A few cocoons or balls with eggs.   Some young spiders recently hatched, if you can find any.   An orb, enlarged, drawn on the blackboard.   In the following lesson the web and habits of the common orb weaver, *Epeira vulgaris*, are described.   Take any orb weaver you happen to collect and describe its color and markings as you find them.   The males of many spiders are much smaller than the females.   Before this lesson the children should observe the three kinds of spiders mentioned, and their habits.

Spiders are not so numerous about our houses as the flies which we studied in our last lesson, but if you collected all the different kinds of spiders you could find, you would be surprised to learn that there are perhaps as many different kinds of spiders as there are different kinds of flies.   To most of us spiders do not seem pretty animals, like butterflies or birds, but the homely spider which I have in this glass had made a nest as wonderful as any bird's nest.

Now let us for a moment examine the spiders we have here. Their bodies consist of two parts: of the abdomen,

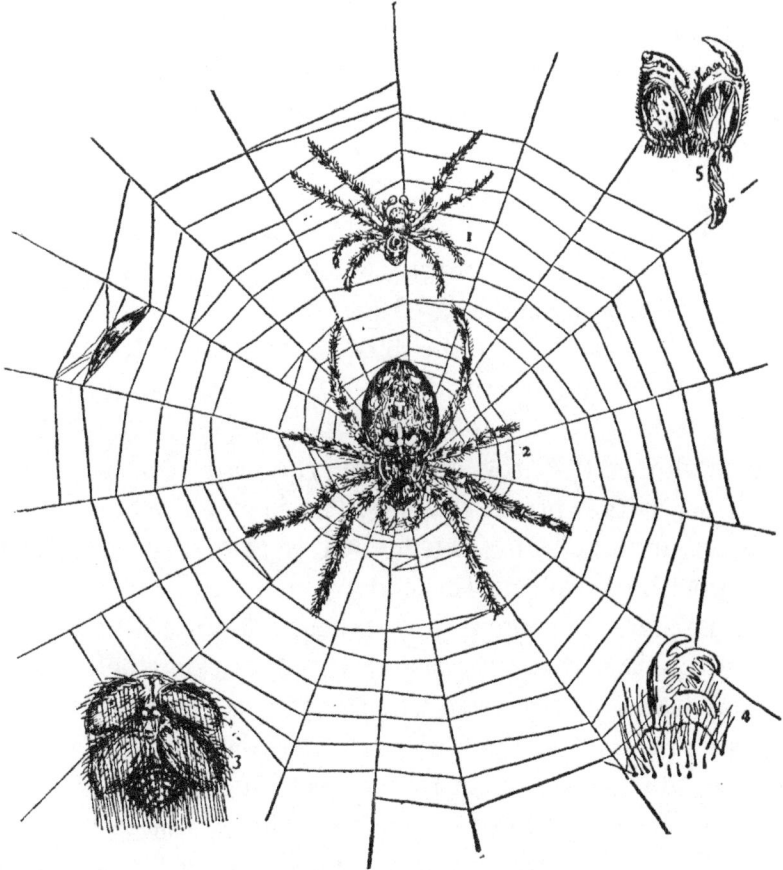

FIG. 21. AN ORB WEAVER.

1, male; 2, female, — both reduced; 3, the spinnerets; 4, a claw of a foot; 5, a claw showing the poison gland. The last three enlarged.

which is soft, and of a hard part to which the eight legs are attached. You remember that the body of a butterfly showed three distinct segments: head, thorax (the middle segment), and abdomen. All spiders have the head and

G

thorax united. If you look closely at the head of one of
your larger spiders, you will discover eight small eyes on it.
In front of the legs is a pair of organs which look like short
legs. They are called palpi, and are used by the spider in
handling its prey. The segments of the palpi next to the
head are flattened, and assist in chewing the food. In front
of the palpi are two strong jaws, each armed with a sharp
claw. Each claw has a small opening near its sharp point,
and when the spider bites, poison from a gland in the head
flows through this opening. The small insects which constitute the spider's prey are quickly killed or benumbed by
this poison, but it has but little effect on the human skin.
There is no well-authenticated case on record where a person
was fatally poisoned by the bite of a spider; but as some
persons are very susceptible to poisons, it is best to avoid
being bitten.

Near the end of the abdomen, you can observe several
small, wart-like protuberances; these are the spinnerets.
Each spinneret is provided with a few large and many
small openings; out of these flows a soft, viscid substance,
which hardens as soon as it is exposed to the air.

Structure of an orb weaver's net. — Comstock in his
Manual says: " Few, if any, of the structures built by lower
animals are more wonderful than the nests of orb-weaving
spiders; but these beautiful objects are so common that
they are often considered hardly worthy of notice. If they
occurred only in some remote corner of the earth, every one
would read of them with interest." The spider first con-
structs a framework of dry and inelastic lines, which are
uniformly taut and quite strong. The outer part of the
framework is irregular, but the central part is regular and
consists of a number of lines which radiate from the centre
of the web. To these radiating lines the spider fastens a
thin, sticky, and elastic thread, usually in the form of

a spiral. Some orb weavers construct nets which differ somewhat from the one just described. Have you ever seen an insect fly against the spider's net? When it touches one of the turns of the sticky line, the viscid thread at once adheres to it. As soon as the insect feels this thread, it begins to struggle in order to free itself; the elastic line stretches and the insect becomes entangled in other turns of it. If the sticky line were not elastic, most insects would simply break the turn or turns they happened to strike, and escape. Teacher and pupils should touch the threads of a web and observe the difference between the thick, supporting lines, and the thin, spiral catch-line.

How a spider seizes and eats its prey. — Some of the orb-weaving spiders live in their nets, hanging head downwards; others have a hiding-place near one corner of the net. They keep hold of some of the lines leading from the net, and as soon as the jar caused by an entrapped insect is thus telegraphed to them, they rush out from their retreat and seize their prey. Sometimes, when the insect is small, they begin at once to suck the blood and moisture out of it; larger insects are often completely wrapped in threads before the spider begins its meals.

Eggs and young. — In fall the orb weavers lay a large number of eggs, which they cover with soft silk, and after hiding the cocoon in a protected place, they watch it until they die, which they do when cold weather sets in. The eggs hatch in the following spring. For a short time, the little spiders remain together in their nest, then they separate, and each spins its own little web.

Other well-known spiders are the Funnel-web weavers, the Cobweb weavers, and the Jumping spiders. The webs of the first can be seen on the grass in large numbers, before the dew has disappeared. Every housewife wages war with broom and brush against the cobweb weavers. You can find jumping and running spiders on any sunny slope very early in spring, and later in the season they are common everywhere. If you desire to study the habits of a spider, put it in a large glass, provide it with a few sticks and straws, and from time to time feed it a few flies. Many birds are very fond of spiders and their eggs. Some wasps kill them by stinging, and then carry them off as food for their larvæ.

NOTE TO TEACHER. — At the close of this chapter a few lessons should be given on the preparations animals and plants make for winter. A few suggestions are given below, but the working out of the details is left to the teacher.

1. Most birds and some bats migrate south.

2. Frogs, toads, snakes, turtles hibernate.

3. Some mammals, like gophers, chipmunks, black bears, and badgers, hibernate also.

4. Weasels, minks, foxes, and wild cats live principally on rabbits and mice, which they can find the year through, therefore they do not hibernate.

5. A few birds that can find food in the winter remain with us.

6. Some insects winter as adults, others as pupas, others as larvæ.

7. All plant life, except some low aquatic forms, is dormant in northern region.

VIII

WINDOW FLOWERS. NOVEMBER

We have studied a few plants that give us food; others that make our gardens and lawns beautiful with their flowers and leaves. Now we will study a few flowers which have gained a place in our homes. Most of the flowers which add so much cheerful beauty to our rooms are children of a warm climate, and cannot endure our winters out of doors. Of the great number of house plants, we can study only a few.

§ 38. The Geraniums. *Pelargonium.*

MATERIAL: Several kinds of geraniums in bloom ; if possible, have also some with seed pods, seedlings, if procurable. Plant several cuttings, as directed below, two or three weeks before this lesson is given. Take them out of the sand and show them to the class.

The home of most of our Geraniums is the Cape Colony, in South Africa. The children there can pick geranium bouquets as easily as you can find goldenrods and sunflowers. The many species have become so much mixed by cultivation, that it is difficult and unnecessary for our purpose to distinguish them.

The flowers vary from white into pink and the different shades of red. The small flowers always grow so close together that they make a conspicuous bunch. Each flower has generally five unequal petals. The so-called double geraniums have more than five petals, a change which has been caused by cultivation. Double flowers have changed

---

Observations. — Is the hair of horses and cattle thin and glossy now as it was in July ?

85

all or some of the stamens into petals. They are often sterile. There are no wild plants which have double flowers.

The leaves are produced in great abundance, and are kidney-shaped, or nearly circular. They are covered with soft hairs on both sides. In some species they show a dark horseshoe band; other kinds have sweet-scented leaves.

Geraniums among the most desirable house plants. — The plants are easily raised from cuttings of almost any length. Geraniums, although they cannot endure our winters, are not very susceptible to changes in temperature, as long as it remains above the freezing-point, and even a light frost does not generally hurt them. If the busy mother forgets to water her plants regularly, some are very much injured; but the geraniums soon raise their drooping leaves, and recover from the effects of a short drought. All these qualities make the geranium a most grateful plant, and enable even the poorest family to adorn their home with a few flowers and green leaves the year round.

§ 39. Remarks and Suggestions on a Few Other House Plants.

The teacher might give a similar lesson on some other common house plant. One or more species of the following can be easily procured almost anywhere: Fuchsia, Begonia, Pink, Cactus, Flowering Maple (Abútilon), English Ivy. Substitute any of the above for the Geranium, if it suits your convenience.

The following remarks may be helpful. Most *Fuchsias* are natives of the mountains in Mexico and South America. The calyx forms a long tube, is attached to the ovary, is brightly colored, and its four lobes stand at right angles to the corolla. The corolla is attached to the calyx. Fuchsias

---

Observations. — Could horses, cattle, and sheep survive the winter in your region without being fed by man?

are easily propagated by seeds or cuttings. The species are numerous and greatly mixed.

Begonias are often cultivated for their beautiful foliage as much as for their flowers. Calyx and corolla are colored alike, and stamens and pistils are found on separate flowers. The plants readily grow from leaves, which are planted edgeways in moist sand.

The Cacti have no leaves; their place is supplied by the green rind of the stem, which is often flattened. By this arrangement, they are admirably adapted to their environment. As they grow in dry places or in arid regions, it is to their advantage to present but little surface to the air, and thus reduce evaporation of moisture to a minimum. A few small species grow wild on rocks and sand in our Northern States, while in the arid regions of our Southwest and West the genus is represented by numerous larger species. As house plants they need but little water. They are propagated by seeds or cuttings.

The Flowering Maple, or Abútilon, has the stamens united in a tube, which surrounds the styles.

The English Ivy (*Hedera Helix*) is a climbing plant adhering to walls by numerous small rootlets. In this part of the United States, it is found only as a house plant. Farther south, and in England and western Europe, it endures the winter out of doors. It is cultivated for its lustrous, dark green foliage.

For either one of these plants, the teacher should provide similar material as for the geraniums.

§ **40.** A Lesson on Window Gardens.

The following directions are condensed from an article by George W. Carver in Bulletin No. 32 of the Iowa Agricultural College, to which the teacher is referred for more detailed information.

Do you know where our horses, cattle, and sheep came from?

MATERIAL: Let each pupil, or as many as possible, be provided with a shallow box or pan. Following the directions below, let each child plant some cuttings and sow some seeds. If there is no danger of freezing, keep as many of the receptacles in the schoolroom as is convenient. Let the children take the others home. After four or five weeks all the plants should be brought to school and the results compared. If seeds cannot be had conveniently, a florist in any larger town will probably send enough to provide a class for twenty-five or fifty cents.

How to raise flowers from cuttings. — All cuttings should be rooted in clean sharp sand (such as plasterers use), the object being to secure perfect drainage, warmth, and aeration. In such soil the cuttings are much less liable to rot and mould.

How to cut. — All plants having conspicuously jointed stems should be cut just below a joint; all others just below the point where the leafstalk joins the stem. It is best to make all cuttings slantingly, because such cut gives more surface for rootlets than a straight cut. Make your cuttings about two inches long, remove all leaves near the lower end, and if the upper leaves are large cut away the upper half.

Receptacles. — A shallow box or pan is a most excellent receptacle for the cuttings. Begonias, heliotropes, fever-fews, lantanas, coleus, carnations, and kindred plants will root best if covered with a large glass, a fruit dish, or a glass tumbler; because the conditions of heat and moisture can thus be kept much more uniform. The sand should be kept constantly moist, and the plants removed as soon as they are rooted. Most cuttings will root in two weeks; with a little care they may be removed from the sand and examined at any time without injury.

Raising plants from seed. — This lesson is important on account of its instructive value to the children. In the

Observations. — Did the Indians north of Mexico have domesticated animals when America was discovered?

case of many plants, seeds will give much finer specimens than cuttings.

All seed should be sown in well-drained, shallow boxes or pans. Use sandy soil which has been passed through a fine sieve.

Smooth the surface carefully; sow the seed and gently firm the soil with the hand or a piece of board. All such seeds as those of geranium, mignonette, asters, chrysanthemums, should have a light covering of soil sieved over them · just enough to cover is all that is required.

Very fine seed, such as that of begonias and tydeas, should be sown and firmed as directed above, but no after covering should be given. Place a piece of blotting paper or soft carpet paper directly upon the earth and water it through this paper, which should be removed as soon as the seeds germinate. Remove the seedlings, and plant them in separate pots or boxes as soon as they have from four to six leaves.

The following plants may be grown with ease by observing the above precautions: Geraniums, oxalis, cinerarias, abutilons, chrysanthemums, coleus, cyclamen (Alpine violet), amaryllis, freesias, and callas.

If the seedlings or cuttings should be attacked by moulds, the following treatment is recommended: —

(a) Dry the plants, leaving just enough moisture to prevent withering; (b) water in the morning only; (c) lower the temperature to the minimum necessary for growth; (d) give plenty of fresh air and sunshine; (e) sprinkle over the soil equal parts of finely powdered sand and flowers of sulphur.

Flowers of sulphur can be bought in any drug store.

If you are interested in the raising of trees, shrubbery, and flowers, see Bulletin No. 34, Iowa Agricultural Experiment Station. Article: Home Propagation.

Which of our domestic animals do you consider the most intelligent?

IX

REVIEW OF PLANTS STUDIED

PARTS OF PLANTS. NOVEMBER

MATERIAL: Plants with different stems from the school herbarium; different kinds of leaves and fruits.

§ **41.** 1. *Roots.* —*By means of roots, plants are held firmly in the soil, but they also furnish the necessary water to the plants, and with it certain food materials.* The roots of trees penetrate very deep into the soil, and they also spread very far all around the tree. How could you find the depth and distance to which roots of trees grow? What different shapes of roots have you observed? Many plants have an underground creeping rootstock, from which one or more plants grow in spring. Mention some such plants.

2. *The Stem.* —*It bears the branches, twigs, leaves, flowers, and fruit.* Most plants have a more or less erect stem. The cucumber has a long, creeping stem. The pea clings to sticks or to other plants by means of fine tendrils, which are wound around their support like springs. Do you know of plants which twine around poles and trees, or climb up walls and trees, holding themselves by means of short rootlets or discs? Examine some Virginia Creeper and Wild Hops.

Many plants can readily be increased by cuttings made from their twigs. How can the wild and undesirable plants be improved by budding and grafting?

Observations. — Which is the most useful?

3. *Leaves. — The leaves are as important to the life of plants as roots.* The largest amount of the food of plants is derived from the air by means of the leaves. If a plant is defoliated, it may reproduce a new set of leaves from reserve material stored in its tissue, but it is always injured and produces little or no fruit. If defoliation is repeated, the plant will die. Potato beetles and caterpillars often kill plants by defoliation. Leaves, as also the stem and branches, give beauty and characteristic forms to plants. Try to recall a number of the different forms of leaves you have observed. Leaves are nearly always more or less thin and flat, like a piece of paper, because this shape enables them to present the largest possible surface to the air. You will not find leaves that are tuberous like a potato. The leaf veins serve to keep the leaf spread out. In grasses and lilies they run nearly parallel to one another, but in the leaves of most of the plants that we have studied they form a beautiful network. The venation is best seen if you hold a fresh leaf between your eyes and the light. By far the larger number of trees and shrubs in northern countries shed their leaves in the fall. As the temperature of the air in our winters remains below the freezing-point of water for months at a time, the water in the leaves would freeze, growth would be suspended, and the leaves would only furnish a large surface for snow and storms. Our trees and shrubs overcome this difficulty by producing a new foliage in spring, and by shedding it when the season's growth has ceased.

4. *Flowers.* — Those parts of a plant which are specially adapted to the production of seed are called the flowers. The essential parts of a flower are stamens and pistils. It is necessary for the growth of seeds that pollen from the anthers be transferred to the stigmas of the pistils. On the

Are they all entitled to be treated with kindness?

moist surface of the stigma the pollen grains send out small tubes, which grow through the style into the tiny ovules (seedlets) below. Unless that union takes place, the ovule does not develop into a seed, but dies. Some flowers, like the tulip and lilies, have only one kind of floral envelop, which is called the perianth. Others, like the grasses, are destitute of both calyx and corolla. Their stamens and pistils are protected by small chaff-like glumes or bracts.

The chief service the corolla or the perianth renders the plant is the attraction of insects. They come to eat the honey secreted by the nectaries, and the showy corolla or perianth act like large bright signs, showing them where to go. In some flowers, as the Fuchsias, the calyx also has a bright color, and serves to attract insects. *The important service insects render to plants is the carrying of pollen from one flower to another.* Inconspicuous flowers are either wind-fertilized or self-fertilized. Flowers which are fertilized by the wind, like poplars, hazels, birches, maples, grasses, etc., produce a large amount of pollen. Can you tell why that should be necessary? Most of our well-known plants have the stamens and pistils in the same flower; some, like the cucumber family, have them in different flowers; others, like the poplars and willows, produce staminate flowers on one tree and pistillate flowers on another. Only pistillate flowers can produce seeds; staminate flowers wither and dry up when they have shed their pollen.

5. *Fruit.* — *The ovary with the mature seeds enclosed is called the fruit.* In some plants the calyx tube enlarges and also forms a part of the fruit. The *Berry*, as shown in the currant, the gooseberry, and the grape, is one of the most common kinds of fruit. Here the whole ovary has become fleshy. The Gourd fruit is simply a very large berry with

a hard rind and a soft flesh within. In the apple, pear, and quince, the enlarged calyx tube forms the edible part. On the side opposite the stem the lobes of the calyx can still be seen in the ripe fruit. In the Stone fruits, the outer part of the ovary has become fleshy and the inner part hard and stony. What other varieties of fruits have you observed? (See Gray: "Lessons in Botany.") It is very important to the life of plants that their seeds should not all remain near the parent plant. Often they could not grow there at all, and if they were not carried to other places, the species would be in danger of becoming extinct. When we studied the common road weeds, we learned that small seeds are often carried for miles by the wind and also by water. You have noticed that nearly all berries, when ripe, change their color from an inconspicuous green to a conspicuous red, black, blue, yellow, or white. Generally the taste also changes from sour to sweet. The conspicuous color enables birds to find the berries, and the sweetness makes them palatable. · The hard seeds in the berries pass through the bird without being digested and grow if they happen to be dropped in a suitable place. Can you think of other contrivances by which plants disseminate their seeds?

X

DOMESTIC ANIMALS. DECEMBER TO FEBRUARY

Of the many thousands of wild animals but few have proved profitable for man to domesticate. How important these dumb creatures have been in the progress of civilization, we shall see in the course of our lessons. Some provide us with food and clothing; others carry burdens or draw heavy loads; a few are kept for the pleasure they give us. Did it ever occur to you that we should, in all probability, still be savages if our ancestors had not tamed and domesticated some such animals as horses and cattle? No native American tribe or nation possessed a domestic animal as strong as horses, mules, or oxen.

§ 42. The Dog.

MATERIAL: The following will be found helpful, if they can be procured: Pictures of different varieties of dogs; a clean skull, to show the dentition. Previous observations: Teeth of dogs; their food; what dogs can learn; their watchfulness, faithfulness; difference in individual characters; dogs drawing sleds and carts. Pupils may supply incidents illustrating the intelligence of animals, but the teacher must not let them side-track her.

Nearly all our domestic animals have been tamed and domesticated by man in prehistoric times. All evidence we can procure shows that the dog was the first animal helpmate and companion to man. In the stone age, when man

---

Observations.—Learn to recognize the following trees in their winter condition:

94

had not yet learned to work the common metals, but made his axes, spear and arrow heads of flint, he had already domesticated the dog. The ancestral wild form of the dog is not known. It is probable that it no longer exists in a wild state, but it must have been an animal much like our wolves and the jackals of Asia.

Why the dog was easily domesticated. — People who live by hunting often capture young animals and take them home as pets. Wild pups caught in this way were easily fed by savages on the remnants of their own meals. As dogs follow their master without any trouble on the master's part savage tribes found their dogs a very convenient food supply, when game was scarce. Some Indian tribes, Esquimaux and the savages of Australasia are as fond of dog flesh to-day as we are of mutton, beef, and pork.

Every country has its dogs. — As the dogs learned to eat almost everything which their masters ate, they were enabled to follow man into every climate from Greenland to New Zealand. To this very day dogs are the only domestic animals of the Esquimaux in Greenland. They draw his sleds, furnish him warm skins, and also meat.

Intelligence of the dog. — That the dog is the most intelligent of our domestic animals hardly needs proof. He not only knows his master, but also a number of other persons. Among his human acquaintances are some whom he likes and others whom he hates. Of his master's praise or blame he shows a very keen perception. If he is caught blundering and laughed at, all his actions show that he feels ashamed. But it is by his conception of property, and by his devotion to his master, that he rises high above other animals. Some other animals can be taught to follow the

---

Willows, aspen, cottonwood, white elm, box elder, soft or silver maple, sugar maple, linden or basswood, hackberry, scarlet oak, bur oak, canoe birch, white ash, butternut, bitternut or swamp hickory, ironwood.

person who feeds them and cares for them; they also show a certain amount of love and devotion towards him; but none except the dog have any idea about their master's property. A good dog will follow none but his master or a good acquaintance; he will not permit a stranger to take or even touch any of his master's property. He learns to recognize the extent of a lot or yard, and even the boundary of a small farm. If a stranger attempts to drive him away from his own territory, he shows his teeth in reply; but if not on that territory, he generally runs at anybody's order. He even seems to know that small children are not quite responsible for their acts, as he will endure much cruelty on their part which he resents if offered by grown persons. Many a dog will rush to assist his master if the latter is assaulted, and almost everybody knows some well-authenticated story where a dog saved the life of a person. Have dogs good memories? Compare the intelligence of the horse with that of the dog.

The senses of hearing and of smell are very acute in the dog. That dogs can follow the tracks of animals and men by scent alone seems almost incredible to us whose sense of smell is not very keen. Their sense of sight is also well developed, but as they do not stand as high as a man, objects which we plainly see are often hidden from their view.

A dog has three kinds of teeth. His front teeth are called incisors. They are rather small; he uses them to peel off flesh from bones. His eyeteeth are very long, and with them he can hold an animal and inflict fearful wounds. The eyeteeth are also called canines. Behind the canines are the molars. These are very strong, and in chewing they cut or crush the food. His jaws are worked by

Observations. — Where can you find hazel, red osier, sumach, wild grapevine, Virginia creeper, false bittersweet, wahoo or burning bush.

powerful muscles, which enable him to crush large and hard bones. Compare the teeth of a sheep with those of a dog. Procure the clean jawbone of a sheep for this com-

FIG. 22. SKULL OF DOG.

parison. Can you judge of an animal's food by its teeth? Foxes and wolves have teeth like a dog. Of what use are the large canines to them? Why should the dog have the large pointed canines?

§ 43. The Cat.

MATERIAL: The clean skull of a cat or a somewhat enlarged drawing of it on the blackboard. Previous observations: Retractile claws of the cat; its teeth; walking, climbing, jumping; attitude against dogs; love for its kittens.

Although the cat was, no doubt, domesticated much later than the dog, it has nevertheless lived in the company of man for thousands of years. It is probable that it was first tamed by the ancient Egyptians and that the wildcat of Egypt and Nubia is its ancestral form. In Europe it did not become common until after the Crusades, when the increased cultivation of grain made it very useful for checking the increasing numbers of mice and rats.

On the twigs of the last two you will find beautiful red fruit.

H

Both cat and dog, as their teeth show, belong to the order of flesh-eating animals, but in many respects they differ markedly from each other. While the dog becomes so strongly attached to man himself, the cat is merely attached to the farm or dwelling-place of its owner. If a family in town moves, they have to catch their cat and carry it with them or it will stay in the deserted home for a while and then try to find another home. In parts of New England where many farms have been deserted, the cats have remained around the old homesteads, and have become entirely wild. Even in thickly settled districts a cat sometimes returns to the wilderness, at least for the summer months. A number of years ago I observed in southern Minnesota a cat which lived in the woods for several years. Only at the approach of a snowstorm or during very cold weather it sought food and shelter at the neighboring farms and occasionally came into the houses with the other cats.

If a dog loses his master, he generally follows the first man who speaks a kind word to him. No cases are known where dogs have become entirely wild and have multiplied in the feralized state.

In intelligence cats are almost the equal of dogs. They easily learn to open a door or to understand other mechanical contrivances; but they will not learn to serve man as the dog does in many ways. The love some people show for their cats is a rather one-sided affection; as it is, at least, very doubtful if a cat has any real love for its master or mistress.

Structure of the cat. — If we compare the movements of a cat with those of a dog, we must call the dog a clumsy animal, while a cat is one of the most lithe and agile animals on the face of the earth. One must see two cats fight to

---

Observations. — After the snow has melted, look for shrubs and young trees whose bark has been gnawed off by rabbits and smaller rodents.

appreciate their quickness of motion. The bones of a cat's leg are joined at angles, which give much spring when the legs are stretched out, and this arrangement, together with the powerful muscles by which the bones are moved, give the cat the power of making the well-known long and sudden springs upon its prey. Both dogs and cats walk on tiptoe, but the cat's walk is much more elastic than that of the dog. The speed of a cat in running and walking is not great, but she does not capture her prey by chasing it. The nails of a dog are always shown; they touch the ground and are dull; but the cat can retract hers into a sheath. She does not walk on them, but always keeps them sharp for catching and holding her prey, for climbing, and for fighting. The teeth of a cat are very much like those of a dog, but her molars are sharper. A dog's tongue is soft above, but the cat's is covered with horny projections, by means of which she can lick the flesh from bones.

Senses. — Cats have keen senses of sight, hearing, and smell. The pupil of their eyes is so constructed that it contracts in bright light, but enlarges in the evening to admit what little light there is. The cat's whiskers serve her as organs of touch. With their aid she feels her way noiselessly through brush and grass even in the darkest night. Thus equipped with sharp teeth and claws, with keen senses, and a noiseless tread, she becomes an ideal hunter. But with these qualities she combines great cunning and patience, and, if necessary, she will lie in wait near a mouse hole for hours.

Courage and affection for her kittens. — With her cousin, the dog, she does not always live on good terms. When we say that brothers and sisters live like cats and dogs, we do not mean to pay them a compliment. When a dog attacks the cat, she generally runs, or climbs a tree; but

---

Notice the characteristic mode of branching of each tree.

if she has kittens, she will attack any dog, almost always driving him off.

Wild members of the cat family. — If you have ever seen lions, tigers, and panthers in a circus, you must have noticed at once that they are simply gigantic wildcats. Look at their claws and teeth, if you have an opportunity. They all catch their prey by crawling up to it and then seizing it with a frightful leap. All members of the cat family hunt singly; but wolves and jackals often hunt in packs. The largest member of the cat family in the United States is the Puma or American Panther.

Sheep and Cattle

MATERIAL : Pictures of different breeds of sheep and cattle ; pictures of wild sheep and cattle would also be desirable. A clean skull of a sheep is very desirable ; a jawbone or several teeth of a cow ; horns of sheep and cattle ; antlers of a deer.

§ 44. The Sheep.

Ancestors and time of domestication. — As early as 3000 B.C. the sheep was a domesticated animal in Egypt. How long before that man first tamed the sheep and kept herds of them is not known. It was one of the domesticated animals of the old Hebrews, as numerous references to it in the Old Testament show us.

Asiatic peoples probably derived their stock of sheep from wild species of that continent. According to recent investigations and experiments made in the "Tiergarten" at Halle, Germany, the European breeds have for their ancestor the Mouflon, which, to this day, lives wild on the mountains of Corsica. Our American domestic sheep have, of course, been introduced from Europe.

The sheep is most valuable to people who need warm clothing for cold winters. You know that savages wear skins in cold weather; but where would all the skins come from to clothe the inhabitants of the northern United States and of northern Europe? Sheep thrive best in a dry and cool climate. Cattle often go into a river or lake to keep cool, but sheep never do that. Can you think of a reason why sheep dislike heat and dampness? All the known wild sheep of the earth live on the wildest cliffs of high mountains, where the winters are very long and cold. One of the noblest wild sheep is the Rocky Mountain Bighorn, of which several small herds live in the Yellowstone Park. It is much larger than our domestic sheep, and the horns of the ram grow to an enormous size and acquire a weight of fifty pounds.

Our domestic sheep is not famous for its intellectual powers; on the contrary, its stupidity has become proverbial. Nevertheless its ancestor, the Mouflon of Corsica, our bighorn, and other wild sheep are the most wary animals and are more difficult to approach than deer and elk. Can you tell how the tame sheep may have lost its wits?

As sheep can crop very short grass and need but little water, sheep raising often pays in arid regions, where cattle raising would not be profitable. Some breeds are principally raised for mutton and others more for their wool.

Although even the largest sheep remain very much smaller than cattle, the two animals resemble each other very much in structure; so that we can study the structure of the sheep in our next lesson, in which we shall learn about tame and wild cattle.

§ 45. Cattle.

MATERIAL: The same as for the sheep; foot of a sheep; pieces of tripe bought at the meat market. Previously observed: Cud-chewing of cattle; their food; how they walk and run; how they defend themselves; cow and calf; sociability of cattle; oxen in the yoke; how flies and mosquitoes molest them.

From written and unwritten evidence, we know that man first tamed smaller animals, such as the dog, the pig, and the goat. Dogs and pigs are omnivorous and required therefore very little care; the goat is at least omnivorous as far as plants are concerned. A much more important step forward was made when the domestication of such large animals as cattle and horses was accomplished.

Several species of wild cattle lived in Europe and Asia in prehistoric times, and furnished, no doubt, the most important large game for primitive man. All species of wild cattle are gregarious; that is, they live in large herds, in which the stronger males are the leaders. Man probably first tried simply to keep these wild herds together and retain them within the territory of his tribe. Thus the first step was taken to tame and domesticate these large beasts. Animals captured as calves were no doubt the first ones entirely tamed and domesticated. These tamed beasts would naturally be used to carry or drag some of the property, when their masters sought another camping place. After a tribe had become permanently settled and was compelled to depend partly on agriculture for its food supply, it was an easy step to yoke their pack oxen to a rudely constructed plough. The advantage of a people who possessed domesticated cattle over those who had none can hardly be overestimated. The former were in possession of an almost inexhaustible food supply, and on the backs of their beasts or on rude wagons they could move their families and their property to any desired place. At first, man did not use the milk of his beasts, and he, no doubt, first learned to milk smaller animals like the goat; but when he did learn to procure and use the milk of cattle, another large food supply was added to his dietary. History teaches us that all nations who did not have some large, strong beast of burden never attained to any high degree of civilization. Our own

Indians, for instance, never tamed the buffalo, and remained savages.

The ancestors of our domestic cattle are probably all extinct now. At the time of the Romans and even late into the Middle Ages the Aurochs roamed through the forests of Europe. Only one herd of a few hundred is now left in a forest of Lithuania. Of the thousands of Bison which less than fifty years ago roamed over our Western prairies only a small herd of not more than fifty individuals survives in the Yellowstone Park. Do you not think it is a shame that these noble animals were so brutally slaughtered?

Return to wild state. — If tame cattle are left to themselves in the woods or on the prairie, they soon turn wild again, and, if the climate is not too severe, increase very rapidly Thousands of wild cattle, whose domesticated ancestors were brought from Europe by the Spaniards, now roam over the plains of South America. The ranch cattle of our Western States and of Texas are half wild. You must read an account of the wild South American cattle and also of the ranch cattle in the United States. Sheep will not increase if left to themselves. Can you tell why not? All wild and half-wild cattle develop long horns and easily defend themselves and their calves against wolves and other beasts of prey.

Food. — The principal food of tame and wild cattle consists of different kinds of grasses; but tame cattle are very fond of grain and also learn to relish various cultivated herbs and roots. Deer, goats, and sheep are very fond of browsing on the twigs of trees and shrubs; cattle, on the other hand, will browse but little if they have plenty of good grass. (Compare the lesson on the grasses.)

Teeth and stomach. — What a difference there is between the teeth of a cat and those of a sheep! The teeth of a cow are like those of a sheep, only much larger. The upper jaws of both sheep and cattle have no front teeth; especially developed eyeteeth, like the canines in the cats and dogs, are wanting in both upper and lower jaws. Notice the large gap between the lower incisors and the molars. What large surfaces the molars have, and what curious hard ridges they show! Why these peculiar teeth? Let us consider how and what the animal eats, and we can perhaps answer the question. Cattle cannot bite off the grass they eat. Why

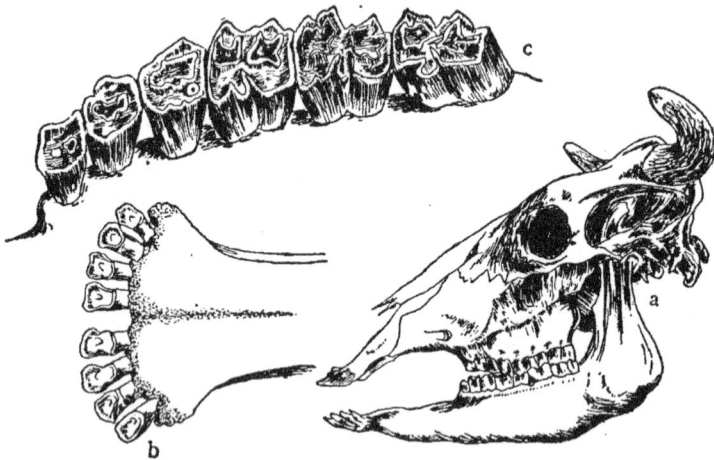

Fig. 23.

a, skull of a cow showing dentition; b, incisors; c, molars.

not? They gather it with their tongue, draw it in between their lower front teeth and the hard upper gum, and then pluck it off. Do they pull it upward or downward? They gather a big mouthful, and, rolling it into a ball without chewing it thoroughly, swallow it. Now you must look at the drawing of a cow's stomach. (See figure.) From the gullet, the food passes into the large paunch to the left; after it has been soaked here for some time, it passes into the honeycomb bag, so called from the cells on its interior wall. Now you have all observed that cattle and sheep often chew and chew for hours while they are lying down, and apparently not eating anything. If you have formed the habit of observing carefully, you must have seen that from time to time they swallow a mouthful, and that this food passes down through the gullet into the stomach. A few moments later, you will notice something passing up through the gullet and into the animal's mouth with a peculiar belching sound. Evidently the animal is far from being sick or uncomfortable, for it at once begins to chew this ball with apparent satisfaction. The ball which you saw passing up is called the cud. It came out of the honeycomb bag. After the cud is thoroughly masticated, it is swallowed a second time; but now passes into the third stomach, which is provided with many folds of delicate membrane resembling the leaves of a book; from this it is carried into the fourth stomach, where most of it is digested, and the remainder passed on into the small intestine. (The teacher ought to observe these parts in a freshly killed sheep.) The whole stomach then, as we learned, consists of four parts: the large paunch, the honeycomb bag, the many-plies or third stomach, and the fourth stomach.

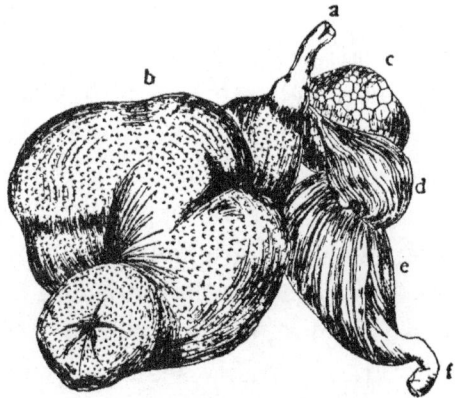

Fig. 24. Stomach of a Sheep.

a, gullet; b, paunch; c, honeycomb bag; d, many-plies; e, fourth stomach; f, small intestine.

How the stomach of the cud-chewers is adapted to their food, size, and mode of life. — I suppose you know that sheep and goats, as well as cattle, chew the cud. All wild cattle, sheep, goats, antelopes, and deer are cud-chewers, and they also have teeth like sheep and cattle. As you see now, many very large and heavy animals are Ruminants (cud-chewers).

All eat nothing but vegetable food, of which all, especially the largest, need a great deal. In order that the animal may completely digest the often hard and tough grasses, leaves, twigs, etc., it has a stomach which first soaks and softens the coarse food, then returns it to the mouth to be *ground* fine (Have you observed the grinding motion and sound of the cow's jaws?), then soaks it again in the many-plies, and finally digests it in the fourth stomach. If the ruminants had to masticate their food at once, the domesti-cated ones would have to walk about all day; but now they gather their food in the cool hours of evening and morning, and later in the day lie down in a cool shady place to ruminate and rest; provided their owner has had sense and sympathy enough to provide them such a place. Wild ruminants, like deer, elk, and moose, must often graze in open places where they are exposed to many dangers; they are able to gather their food in a short time and then retreat into a thicket, or lie down in the tall grass and brush, where they are not easily seen and where flies cannot molest them so much.

Adaptation of the legs and feet of ruminants to their en-vironment. — That the heavy body of ruminants has strong legs to support it, we see every day; but it is the animal's foot which shows the most beautiful adaptation to the needs of the animal. Cattle in their wild and half-wild state often must make long marches over hard and stony ground to find water and food. Domestic cattle move about in their pastures and are employed to draw ploughs and wagons. Deer and antelopes travel long distances in search of food and water, and are often compelled to flee from pursuing enemies. For these uses the cloven foot answers the purpose exactly. The two strongly developed toes are encased in a thick horn shoe, which wears but slowly on the hard prairie or in the brush. The animals need no shoemaker, and if they run about do not need to trim their hoofs, because they grow exactly as fast as they wear away. On the back of the foot are two small toes, which do not touch the ground when the animal walks on a hard surface, but they are of use in getting over swamps and bogs. I think you can see now that an ox would soon have sore feet if he had to walk on four bare toes, like a cat. Why must the hoofs of cattle be trimmed if the animals are entirely stable-fed?

Weapons of self-defence. — If you make a dog angry, he will try to bite you. The teeth with which he tears and crushes his food are also his weapons of defence and attack. Cattle use their teeth only to masticate their food. When a cow defends herself or her calf against a dog, or fights with other cattle, she uses her horns, and every dog knows that they are formidable weapons and avoids them carefully. Most bulls become vicious when they are about four years old. Many a farmer has been gored and trampled to death by his own bull. This shows that the males of our cattle still retain a good deal of their wild fury, although the species were domesticated thousands of years ago. Do you think horns are of any use to domestic cattle and their owners? Are they not a nuisance? Why do some farmers dehorn their cattle? The horn itself is hollow, but it fits tightly over a bony core, which contains many blood-vessels. Does it hurt the cow to cut the tips of her horns, if you do not touch the core? Do you think dehorning cattle is cruel? It causes them considerable pain. Some domestic cattle have no horns.

All wild cattle, sheep, and antelopes have horns, which they need very much to defend themselves and their young against beasts of prey; but they also use their feet for this purpose. Ruminants of the deer family bear branched antlers, which on our noble elk and moose reach an enormous size, attaining a weight of from fifty to eighty pounds.

And these antlers grow in about ten weeks. Every year they are shed towards the end of winter, and they grow anew in summer. The bulls and bucks only have horns, with which they fight fierce battles against each other. Against beasts of prey, the does and cows of this family strike with their feet. But we must postpone learning more about these most beautiful of all mammals to some other time.

Products derived from cattle, sheep, and other ruminants: —

As the facts concerning this part of the lesson are well known, only an outline is given here, which the teacher should expand and treat of in detail.

1. Meat — beef, mutton, venison.
2. Milk, butter, and cheese.
3. Hides for leather.
4. Wool for clothing.
5. Hair for felt, etc.
6. Tallow.
7. Valuable manure.
8. Bones and horns.

The teacher should read some good account of the transportation of live-stock from this country to Europe, and about exportation of frozen mutton from Australia, and tell the pupils about it.

Formerly oxen were used as draft animals much more than now; they were thus used by the early settlers on our prairies. What makes oxen valuable draft animals in a new country? Why are they almost entirely supplanted by horses when the country becomes well settled?

Conclusion. — *We learned some time ago why the grasses are the most valuable plants. In the lessons just finished we have learned that the cud-chewing animals provide us with a large and very important part of our animal food, and that with their wool, skins, and hair they furnish us the most valu-*

able material for clothing and foot wear. Do you think we could be very comfortable without cattle, sheep, or goats ? To this order also belong most of the Big Game animals of this and other countries. But nowhere in the temperate zone is there a fleeter runner than our Virginia Deer, larger game than our Bison, a more majestic creature than our Elk, or a grander Monarch of the Wilderness than our Moose. Every patriotic citizen ought to do his best to preserve these grand beauties of our forests and mountains. Unless the laws regulating the hunting of them are conscientiously observed, all of these beautiful animals are in danger of becoming forever extinct.

Send to the Secretary of the Interior for a report on the Yellowstone Park. Read Shaler's book on Domesticated Animals.

§ 46. The Hog.

MATERIAL : Pictures of wild boar, and of one or more breeds of domestic hog; a clean skull or jawbone, which can be obtained in any household or from any butcher. Previously observed : Rooting, blowing, grunting, and squealing of hogs ; rolling in mud ; how other hogs are attracted by the squealing of one.

Although the pig is generally not very highly spoken of, it has nevertheless performed, and still performs, an important duty for man. It furnishes savory roasts and sausages for millions of hungry mouths. As soon as a tribe of savages becomes settled, their supply of meat has to be largely obtained from domestic animals. For this purpose man tamed an animal which largely supplied his wants, and which has never been raised for any other purpose. Our pigs are descended from the wild hogs of Europe and Asia. The wild boar still lives in its native state in some of the larger forests of continental Europe; but in most places the creatures have been exterminated, because they do very much damage in fields which adjoin the forests. The domestication took place at a very early period, and must have been quite easy, because the animal is truly omnivorous. When the Israelites left Egypt and when the Greeks laid siege to ancient Troy, the hog was a common domestic animal in southwestern Asia and in central and southern Europe.

Habits. — The pig is not held up to children as a model of cleanliness. It likes to lie down in muddy pools; it eats many things which we do not consider clean, but so do chickens, ducks, and geese, and yet we eat them all. Poor piggy has no long bushy tail with which to drive away the

FIG. 25. EUROPEAN WILD BOAR.

flies, nor is he so built that he can reach every part of his body with his teeth. Therefore, when the thermometer registers eighty in the shade and flies are thick, he rolls and dozes in the mud or pool, and thus he solves the heat and insect problem at once. Too often man compels the pigs to be very much dirtier than they like to be. If pigs have a clean place to sleep in, most of them will keep their bodies cleaner than horses keep theirs.

Intelligence and sympathy. — Every farmer knows how quick pigs learn to recognize the person who brings them food, and in how many ways they learn to find food for themselves. It is very easy to make a pet out of a pig, so that it will follow its master everywhere, and pigs have been taught to do things which a horse would scarcely learn. Had man at all tried to develop the natural intelligence of these animals by careful breeding, our pigs might be as intelligent now as our best dogs. I have personally witnessed the following instance of our porker's wits. When a boy, I was bidden with several companions to keep the pigs off a certain meadow. Our task was not an arduous one; for the pigs had to cross, on a single plank, a creek, which was about ten feet wide and had high banks and a muddy bottom. But one old sow had already discovered this secret path, and it was not long before we saw her feeding on the juicy grass of the forbidden meadow. Three or four of us, all provided with whips and switches, crossed the creek. We had a plan to catch the old thing and "to give it to her." After we had turned the plank on edge, we approached the animal as near as possible and then rushed at her. Being of the common, fleet-footed breed of the country, she easily outran us. All of us were much amused at the predicament in which we had the old grunter; we were sure she would stop at the creek. When she reached the plank, she did stop and, for an instant, looked puzzled; but in much less time than it takes to tell it, she had solved the problem. With one quick push of her snout she turned the plank on its flat side and ran across. Although we were not more than about fifteen feet behind the animal, not one of us had a chance to use his switch, so quick was the hog to see through our contrivance. I must say that the surprise was nearly all on our side, and that it took some time before we realized that a group of "smart" boys had been simply outwitted by a sow. *Pigs show a great deal of sympathy with their own kind.* If one utters the characteristic cry of distress, others will at once rush to its assistance and attack its assailant. In this respect they seem to rank even above the dog. What else can a pig indicate by its voice, besides distress?

Value of hogs to man. — As far as meat alone is concerned, the hog is at least as valuable to us as horned cattle. They attain their full growth in a much shorter time than cattle; their meat is easily preserved by salting and smoking, and can be cooked in its own fat. For men who do hard physical work, as farmers, lumbermen, miners, ·soldiers in the field, and sailors, pork is the principal meat diet. In the big slaughterhouses of Chicago alone from thirty to sixty thousand grunters are converted into pork in one day. Among modern nations, Americans and Germans are the greatest consumers of pork.

Structure. — The feet of the pig are cloven; each foot has four toes, but only two touch the ground. Although the foot of a pig resembles a cow's foot very much, the pig's dentition and its stomach are entirely different from those of cattle. You see at once, if you examine a pig's jaw-bones, that it has all three kinds of teeth. In the wild boar, the eyeteeth are powerful, long tusks, which are curved upwards and project from the mouth. The boar cannot chew with them. Can you think what use he may make of them? In the domestic pig, these teeth remain small. With its front teeth the pig can graze, shell corn, pick up nuts, acorns, worms, and other food. Now let us look at its molars. Do you think they are fit for grinding, like the molars of a cow? Why not? Are they sharp and pointed like the teeth of a cat? Do hogs eat flesh? Do they move their jaws like cattle and horses, when they eat?

A hog's stomach is a very simple affair. If you have never seen one, look for it the next time a pig is killed in

the household, or ask the local butcher to show you one.
It consists of only one compartment; the digestion of the
food begins at once, and none of it is masticated a second
time. Although the animal's digestion is very strong, piggy
does not eat hay and straw, because he knows that he could
not digest it and that it would make him sick. In this respect he surpasses in wisdom many little folks and some
big folks. What is the difference between a hog's and a
cow's stomach? Can you tell why a hog could not digest
hay and straw?

NOTE. — Hog's meat should never be eaten raw in any form. So-
called measly meat causes the tapeworm which lives in the human
intestines and often proves very troublesome. A still more dangerous
parasite is the trichina, a very small worm. If infected pork is eaten
raw, the young trichinas find their way from the stomach into the
muscles, where they remain permanently at rest in little capsules.
If present in large numbers, they cause serious illness, and they may
even cause death.

§ 47. The Horse.

MATERIAL: Pictures of different breeds of horses. A clean jawbone, hoof, and bones of a foot would be valuable. Previously observed: Bearing of different horses, biting, kicking, grazing.

The domestication of the horse was accomplished in prehistoric times; but it was most likely one of the last animals
subjugated by man. A high civilization had been main-
tained in Egypt before the horse was introduced by invad-
ing nomads from Asia. The patriarchs of Israel, although
they were nomads, had no horses; and it was not until the
time of Solomon that the horse became common among the
Israelites. There are still wild horses living in central
Asia. They are of a mouse-gray color and about the size of
ponies. Whether these wild horses or some extinct forms
are ancestors of our horses is not known. No wild horses
lived in America when the country was discovered by

Columbus. Where do the wild horses in South America come from?

Value of the horse to civilization. — A number of qualities combined have made the horse the *most valuable* of all our domestic animals.

1. *It is of just the right size.* — If it were as large as the elephant, it could not be properly controlled; if it were very much smaller than it is, it would be too weak.

2. *Its body has just the right shape for the rider.* — Compare the horse and the ox as saddle animals.

3. *Although it is a big animal, it can nevertheless travel fast.* — Can you tell why most farmers prefer horses to oxen?

4. *Its foot, covered by one strong hoof, is the best foot in the whole animal world for travelling far over hard ground.*

5. The horn of the hoof is thick enough to admit of an iron shoe being nailed under it, *and this fact adds greatly to the value of the horse.* If unshod, its feet would soon become sore on our hard, artificial roads and our paved city streets; but with his iron shoes on, a horse can walk all day on the hardest paved street. Horses that run free in the pasture do not need to be shod. Can you tell why not?

It is a curious fact that the ancient Greeks and Romans did not know the horseshoe. It came into use about 300 A.D.; but the inventor is not known.

6. *The peculiar arrangement of the horse's teeth enables its master to control it by means of a bit.*

7. Although it thrives best in moderately warm and moist regions, *it has shown great power of acclimatization* and has spread far north and south. It breeds as well in the domesticated as in the wild state; and its body is so elastic that by a careful selection of breeders, man has produced many varieties (or breeds) suitable for different purposes.

The elephant, although much more intelligent than the

horse, does not regularly breed in captivity. The domesticated stock has to be kept up by animals caught from the wild herds. Is it not strange that these enormous brutes will very soon obey the word and beckon of their masters, who have absolutely no physical control over them?

There is no doubt that horses were at first domesticated for the meat they furnished. The ancient Germans, Anglo-Saxons, and Scandinavians sacrificed horses to their gods and were fond of horseflesh; the value of horses as draft and saddle animals they learned later. In many large European cities, where meat is very high in price, several shops publicly sell horse meat and horse sausages, and they have a considerable trade, but mostly among poor people. The Greek heroes of Homer did not ride to battle on horseback. Horses were employed only to draw their war chariots; in agriculture, the Greeks of that time used oxen.

In modern times the horse has nearly supplanted the ox as a draft animal, and completely supplanted it as a saddle animal.

However, within the last twenty-five years a peculiar revolution has been going on. *The noble horse himself is being supplanted by steam and electricity.* Not more than ten years ago our large cities employed thousands of horses in their street-car systems; to-day there is not a single horse-car in several of them. The result is that horses are very cheap. But in spite of this revolution it seems that farmers, the world over, will always have need of a large number of horses.

The horse in war. — The horse has probably been the most important factor in the aggressive and victorious march of the Aryan race to the conquest of the world. The armed European horseman struck terror into the hearts of the bravest Mexican and Peruvian soldiers, and to-day the horse-less American natives have either vanished or dwindled into insignificance. The Arabs, who were perfect horsemen, conquered northern Africa and Spain within about a hundred years. They only met defeat when they encountered the Franks, who had probably stronger and more enduring horses than they.

At the time of the mediæval knights war could not be carried on without horses. It is true that through the invention of gunpowder the cavalry lost much of its importance. Still, if cavalry is of less importance, the heavy artillery is of so much more importance; and how could a general move his heavy guns about the field if he had no horses; or how could food and ammunition be distributed to millions of soldiers scattered over hundreds of square miles if the army had no horses? Can you think of other work for which an army must have horses? *Thus we see that for purposes of war the horse is absolutely indispensable to us; because it lends to man's intelligence the physical power and the fleetness of its noble body.* A good horse and a good rider almost act as one being.

Intelligence of horses. — One who has carefully studied the subjects, and compared the horse in this respect with the cat and the dog, must come to the same conclusion with Shaler in his interesting book on "Domesticated Animals," that the intelligence of the horse is popularly rated too high. It does not show the real, almost human, affection for its master of which the dog gives unquestioned evidence; nor has it the cunning of cats and monkeys. The insane fright it shows at many strange and even some frequently observed objects does not speak high for its intelligence. Many horses have, however, no doubt a good memory, and will learn a road sooner than most drivers. As their nose is not near the ground, they must follow the road by the objects seen along it. Have breeding horses been selected with a view to rearing intelligent offspring, or to perpetuate some desirable physical quality? On what principles does man select breeders among dogs?

Structure. — The horse has only *one toe* on each foot. Compare a horse's foot with the feet of other animals that you have studied. The first horses which appeared on the earth had four toes, which were separated from each other, and the animals were about as large as foxes. On this interesting subject you must consult some elementary geology. Compare the dentition of the horse with that of mammals you have studied. Does a horse ruminate? Are its teeth fit for grinding hard food? How does it move its jaws when eating? Why can horses graze on very closely cropped meadows?

The Poultry Yard

§ 48. Domestic Chickens.

MATERIAL: Pictures of different breeds; a clean skull prepared by boiling a head with the bill on it. Previously observed: Hens scratching for food and protecting their chicks. For pictures, see Farmers' Bulletin, No. 51, Standard Varieties of Chickens.

The jungle fowl of India is regarded by naturalists as the ancestral type of our domestic chickens.

OUTLINE FOR LESSON

1. *Thoroughly domesticated.* — Do not return to a wild state, are never found far from human habitation. Compare turkeys and guinea fowls.

2. *Food.* — Almost omnivorous; therefore easily kept. Find their food on the farm; almost live on kitchen scraps in town.

3. *Do well if much confined.* — Need but a small grass plot for a run. Compare turkeys and guinea fowls.

4. *Prolific layers and breeders.*

5. *Live in flocks.* — A number of females with one male. Compare robins and other wild birds.

6. *Chicks able to run about,* soon after they have emerged from the eggs, can see, and are covered with a soft.down. Compare the young of robins, sparrows, and swallows.

7. *How the hen protects and defends them.*

8. *Compare* the skull of the chicken with the skull of a cat or sheep. Chicken's bill pointed to pick up seeds and insects; teeth wanting; food swallowed whole.

9. *Feet and wings.* — Feet strong, fit for running, walking, and scratching; wings little used. Compare other birds.

See Farmers' Bulletin, No. 41, Fowls : Care and Feeding.

§ 49. Ducks and Geese.

MATERIAL : Pictures of ducks and geese ; bill of a duck. Previously observed : Walking, flying, swimming, and diving of ducks and geese.

In nearly all countries of the world there are different species of wild ducks and geese. Those tame ducks and geese which were brought to America from Europe are derived from wild species of that continent. Species of the duck family, to which the geese also belong, are very numerous in our country. About forty have been distinguished in the United States and Canada east of the Rocky Mountains, and many a country boy knows half a dozen different kinds of wild ducks. Some of these wild birds, as the Canada Goose, are easily tamed.

OUTLINE FOR LESSON ON THE DUCK

1. *Is slow and awkward on land.* —Compare with chickens. Cannot run well, cannot scratch in the soil for food.

2. *Graceful and at home on the water.* — Body like a round-bottom boat; feet webbed, used as paddles and rudders. Ducks and geese have an oil gland near the tail, take the oil with their bill and oil their feathers, which on that ac-

count are not wetted by water. They shake off any adhering drops after swimming, and are dry. Is this of any value to them? Do chickens get wet in a rain? Do they like it? Do you feel warm or cold when your clothing gets wet?

3. *Dive for aquatic animals and plants.* — Bill covered with a soft skin by means of which they feel about in the mud for their food. Horny plates on bill and fringes on tongue form a sieve to let the water run out and retain the prey captured.

4. *Ducklings take to the water instinctively.* — Have you observed how amazed and worried a hen is when her supposed chicks suddenly tumble into a pond? Hens are often set on duck eggs.

5. *Wild ducks and geese are good flyers.* — Do their tame relatives fly well? Why not?

6. *Wild members of the duck family migrate.* — Recently tamed birds become very restless in fall and spring, and often leave with their wild friends, unless their wings are clipped. Do thoroughly domesticated birds show the migrating instinct?

7. *Show how ducks and geese are adapted to their environment and to the life they lead.*

§ 50. Influence of Man upon Animals.

MATERIAL: Pictures of different breeds of domestic animals and, as far as possible, some of their wild relatives or ancestral types.

As young animals generally closely resemble their parents, man has been able by a careful selection of breeders to raise just such animals as he desired. We have to-day very large dogs and horses, and also very small ones. Shape, color, and intelligence of animals can also be influenced by breeding. Many so-called full-blooded types have been created by man within the last one or two hundred years; a few are older, but some are much younger.

What the dog had to learn, and how its body was changed: —

1. Was originally wild; hunted and killed animals as wolves do to-day.

2. Has become used to man and obeys his word.

3. Hunting dogs have learned to pursue or fetch game without eating or injuring it.

4. Shepherd dogs take care of sheep; must often bite them, but do not seriously injure them or kill them. Do not seize them by the throat as their wild ancestors and relatives did, but bite them in the legs.

5. Watchdogs know their master's family, friends, and property; keep off thieves.

6. Most dogs differ much in size, shape, and color from their ancient prototypes, the wolves and jackals.

How the horse changed, and what he learned: —

1. Originally roamed wild over the plains. Wild and half-wild horses are extremely shy and wary.

2. Had to overcome his fear of man, dogs, etc.

3. Became used to harness, saddle, and bit.

4. Some became much larger, others smaller than their wild prototype; some developed great strength, others great speed, others great endurance.

5. Both dog and horse, but especially the former, have grown in intelligence.

Changes in cattle: —

1. No longer wild and ferocious.

2. Became accustomed to the yoke.

3. Will give their milk to man.

4. Some breeds produce much more milk (Jerseys), others much more beef than wild cattle.

Changes in the pig: —

1. Has lost the extreme wildness of its prototype.

2. Snout much shorter, body more round, legs shorter, fewer and shorter bristles than the wild boar.

3. Produces much more fat and meat.

Traces of the wild state: —

1. Some easily become wild again. Horses, cattle, cats.

2. Males of wild animals are very pugnacious, fight fierce battles with each other for possession of the females. Domestic animals do the same, if they have an opportunity.

3. Most bulls and stallions are still dangerous brutes.

4. If two or three dogs can run about together, their ancient instinct is apt to revive; they go hunting, kill sheep, and commit other mischief.

NOTE. — Read : Shaler, Domesticated Animals.

§ 51. Our Duty to Animals.

All of our domestic animals are derived from wild forms. — Some of the wild ancestral forms are now extinct. Most of our animal helpmates were tamed so long ago that we know little or nothing of that important process. The most recent addition to the list is our American turkey. *Without the help of large, strong animals to do our heavy work, and without the humbler ones who furnish food and clothing for many, many millions of men, we could not have attained our present state of civilization.* If all our domestic animals should disappear suddenly, most of us would probably starve. People who did not domesticate wild animals always remained savages.

Kindness to animals. — As we owe so much to animals, it seems as if all men would naturally treat the dumb creatures with reason and kindness; but to our shame it must be admitted that many a brute's master is far more brutish than the poor beast he abuses. If an animal must be killed, nearly everybody, it is true, does it in a manner which is quickest and least painful to the animal. But often a man's laziness and thoughtlessness inflict much suffering upon animals. Horses are left on the streets unblanketed on a cold winter day for hours at a time. Cattle are compelled to spend the nights in wet, dirty yards; poultry is shut up in foul-smelling coops, which have not been cleaned for years, and are full of vermin besides. The poor hogs are occasionally confined in sties where the mud is a foot deep. All our animals should have free access to reasonably clean water unless it is abundantly furnished in their food. A part of the foul pond water, which a cow is forced to drink, man drinks with that cow's milk. Animals prefer reasonably clean water, but do not like it very cold; and they all thrive better with reasonably clean housing.

It is also brutal to overwork animals without a very good reason. Driving a fire-horse to death in order to save human life and valuable property may be justified; but one who overworks a beast without such good reason should be put behind the bars, where he belongs.

INTRODUCTION

§ **1.** Trees add much beauty to our farms, to cities and parks; their wood is of much use to us for fuel, lumber, and furniture; they furnish the necessary wind-breaks on thousands of prairie farms; they are the homes of the birds, and they shelter small and large game; and, most important of all, they exercise a beneficial influence upon the climate of the whole country. For these reasons they should occupy a prominent place in any plan of Nature Study.

The plan followed here suggests that the trees, common in the neighborhood of the school, be studied three times First, in their winter condition, when the characteristic modes of branching and the forms and arrangement of buds can be best observed; second, when they are in bloom; and third, when leaves and fruits are fully developed. The pupils should be directed to observe the different trees, shrubs, and vines, from very early spring until they again appear in their winter condition. Children, as well as grown people, are easily interested in the doings of animals and plants and in the process of their growth, while a mere description is generally of interest to the specialist only.

The number of trees native to the forests of North

America is very large, much larger than the number of species native to northern Europe. But in addition to our native species, many European trees are frequently found planted in our Eastern States; so that it is often quite difficult to decide which species of tree one has before him, although one need not be in doubt about the genus. If the teacher is in doubt about the species, she should not give a specific name, but should simply study and describe the tree as an elm or a willow, as the case may be; and be sure that the pupils state what they actually see, and not what, according to the teacher's book, they ought to see. If, for instance, the European elm or the slippery elm is chosen instead of the white elm, the description of the latter must be changed to apply to the tree under consideration. In this book the sixth edition of Gray's "Manual" is followed in the nomenclature. That Latin names are given for the benefit of teachers only goes without saying. Gardeners, farmers, and woodmen can generally identify our common trees. However, quite frequently they apply different common names to the same tree; what one calls a red oak, the other is likely to call a black oak.

As far as practicable, trees should be observed in the woods as well as in open places. It must not be forgotten that the color of the bark and twigs, the mode of branching, the shape of the leaves, and especially the size of the tree vary somewhat with local conditions. Many of our common trees attain their largest size in the forests of Ohio, Indiana, and Kentucky, decreasing in size further north and west. Trees which predominate in our Northern forests are not so common in the South. Sometimes they grow there only on the mountains, or are not found at all.

Let the children observe that the snow still remains in the woods, in groves, and under brush, after it has disappeared from the open fields. They must also observe that the snow is not blown out of the woods, and that much snow blown into them is also retained. Why are there no high snow banks in large woods as there are on the prairies and open fields? Show the children how leaves and flowers in the bud are protected by scales from drying up and from too sudden changes of temperature. In January and February the closely packed and folded leaves and flowers in the buds contain but little moisture, and are not at all injured by a steady frost; but after the buds have opened, the young leaves and shoots are filled with sap, and a late spring frost frequently kills every leaf and shoot in whole groves. Have the children verify these statements by observations, and in the same manner learn and teach which trees are injured most by late spring frosts. Of every tree studied preserve winter twigs, flowering branchlets, and full-grown leaves in the school herbarium. Most fruits are best preserved in small boxes. They should be left on their peduncles or stalks, and as nearly in their natural condition as is possible.

The children might be asked to plant the seeds of different trees and shrubs on small plots of ground. All tree seeds, except poplars and willows, which need but very little covering of moist earth, may be planted about an inch deep, in good garden soil, as soon as they begin to fall. You will find, by experiment, that some begin to grow at once, others remain dormant until next spring, and some do not germinate before the second spring. The seed bed may be covered in fall with a little hay or straw, to prevent repeated freezing and thawing. The germination of the seeds and the growth of the baby trees will be of much interest to the children. Unless you have noted the place of the different seeds with marked sticks, you will not at once recognize all the seedlings, so little do many of them look like their giant parents.

In the lessons which follow, only a few trees have been described at length, to suggest how *all should be studied*. Very brief descriptions of the others are given in order to gain space, but the language of these outlines should not be used in the class-room.

After a tree has been studied in its different conditions, the parts of its annual history should be connected into a whole; and when the subject of trees has been completed, different pupils might write out, in the best English they command, the annual history or the life history of different trees.

The following books would be helpful in the description and determination of trees : —

K. C. Davis. Key to the Trees of Minnesota in their Winter Condition.

Knobel. Trees and Shrubs of New England.

Newhall. The Trees of Northeastern America. This includes some naturalized species and all the native trees of Canada and the northern United States east of the Mississippi. The last two books are illustrated, and the key is based on the shape of the leaves.

If you have access to a public library, consult *Sargent*, Silva of North America, the American classic on trees.

II

SOME FOREST TREES IN THEIR WINTER CONDITION. FEBRUARY AND MARCH

§ **2.** MATERIAL : Twigs and a few small branches from every tree to be studied. From a wood pile or from a dealer's wood yard select pieces of wood and bark, which were cut from trunks or large branches. Before these lessons are given, the children must closely observe the shape of the trees in the woods and in open places. If practicable, let them accompany and assist you when you collect the material. Outdoor observations are of the utmost importance in all lessons on trees, because the full-grown plants cannot be brought to the schoolroom. If any of the trees described below are rare in your neighborhood, substitute common forms for them, and omit those of which you cannot, at least, procure branches, twigs, flowers, leaves, and fruit. Put twigs of all trees studied in glasses or bottles with water, and watch the buds.

The Poplars

1. The Quaking Asp. *Populus tremuloides.*

This tree generally grows from thirty to fifty feet high and, a few feet above the ground, the trunk has a diameter of ten to eighteen inches.

Near the base, *the bark* of old trees is often black, from one to two inches thick, and divided into broad, flat ridges. Higher from the ground, and on young trees, it is thinner, pale yellow-green, orange-green, or almost white, and often marked with horizontal, wart-like excrescences. The lower side of large branches is often marked with large black scars.

Observations. — Keep a close lookout for the flowers of all the trees you are studying.

General appearance. — The stem preserves its size with little diminution in diameter for thirty feet or more. At about that height, slender and often contorted branches are produced, which are quite remote from each other. With the straggling, somewhat pendulous, spray into which they end, they form a narrow, round-topped head.

The branchlets are reddish-brown and lustrous the first year; in the second year they are greenish-gray; finally they turn dark gray and nearly all are very much roughened by leaf-scars.

The dark brown buds are covered with a fragrant varnish, and unless already partially developed they are about one-fourth of an inch long and terminate in a rather hard and sharp point. Under each bud a scar is distinctly seen. These scars are the marks left by the falling leaves in autumn. If you take an oak twig and remove the still adhering leaves, fresh leaf-scars can be seen.

Of course, now we desire to know what these buds contain; and in order to find out, we shall make longitudinal sections of some with a sharp knife, and from others we shall carefully remove the brown varnished scales. We find in some large buds a considerable amount of soft, fuzzy scales and a green body which looks like a small catkin. In the smaller buds, John thinks he sees tiny, greenish leaflets which are very closely packed. I am not quite satisfied with the result of this investigation, and think that our inquiries may be answered more satisfactorily if we examine the buds of the elder, the lilac, and the alternate-leaved cornel (*Cornus alternifolia*). Now we need no longer be in doubt about the contents of buds. You have all found very small, closed flowers in some, flowers and leaves in others, and in most of them simply leaves.

Observations. — Note the appearance of the following birds, and continue to observe them.

Range. — This tree has the largest range of all North American trees.   If we could make a long fishing and hunting trip from southern Labrador to Hudson Bay, then northwestward to the lower course of the Mackenzie River and from there to the mighty Yukon in Alaska, we could pitch our tents in a grove of quaking asp almost every night.   If from Alaska we roamed along the mountains to southern California and then across the continent through northern Missouri and through Pennsylvania to the Atlantic coast, we should have made a trail around the enormous territory in which every child may hear the whisper of the aspen leaves.

2. From the following list, select for study with the children the tree you can most conveniently observe.

a. **The Cottonwood.**   *Populus monilifera.*
b. **The Balsam Poplar or Balm of Gilead.**   *Populus balsamifera.*
c. **The Lombardy Poplar.**   *Populus dilatata.*

The latter is the poplar with the familiar spiry top.   It does not grow wild in our forests, but has been introduced from Europe.   Any farmer can point out the Cottonwood and Balsam Poplar to you.

The following topics are suggested : —

> Size and shape.
> Bark of trunk and large branches.
> Where the tree grows.
> Branchlets.
> Buds.
> The wood.

Catbird, brown thrasher, Baltimore oriole, rose-breasted grosbeak, chipping sparrow, yellow warbler, red-eyed vireo.

K

The Maples

3. The Soft Maple, or Silver Maple. *Acer dasycarpum.*

This tree grows to a height of about sixty feet. It generally divides into three or four stout, upright, secondary stems ten to fifteen feet from the ground. These secondary stems are destitute of large branches, but bear a number of brittle, pendulous branchlets

The bark of old trees is reddish-brown, furrowed and separated into large, thin scales; but young trees and large branches have a smooth gray bark, tinged with red.

The smooth and *very lustrous branchlets* of the last season are bright chestnut-brown and dotted with large, pale lenticels.

It is a common tree *along river banks.* In the West it is often planted for shade, and if placed in good, moist soil, grows very rapidly.

The rather large brown flower buds, which appear in thick clusters, can be recognized as early as December.

4. The Sugar Maple, or Rock Maple. *Acer saccharinum.*

A young sugar maple growing in an open place develops a symmetrical oval top of stout and nearly upright branches. As the tree grows older, its branches spread gradually, making a broad, round-topped dome from fifty to seventy feet across.

In the forest, where many trees crowd each other, sugar maples often rise to a height of fifty or sixty feet without a single branch.

The bark is one-half to three-fourths of an inch thick, broken into deep longitudinal furrows, separating into small scales on the surface. On old trees it appears light gray-

Observations. — Preserve some pine flowers in alcohol.

brown or bluish-gray; on young trees it is pale, smooth or only slightly fissured.

The branchlets of the last season are orange-brown, marked with numerous pale lenticels, and quite lustrous.

The winter buds are dull brown, not covered with resin or varnish, and along the sides of branchlets you will always find two buds opposite each other, while a strong bud terminates each branchlet.

Compare the buds of maples with the buds of poplars.

The sugar maple produces the most valuable wood of all the maples. It is used much for flooring and inside finishing, and makes excellent fuel. It is scarcely necessary to state that this is the tree to which boys and girls are indebted for the much-relished maple sugar. Let a boy who has assisted in making maple sugar or maple syrup describe the process to the class.

5. The White Ash. *Fraxinus Americana.*

A large tree, fifty to seventy feet high; with stout, upright, spreading branches, forming a broad, round-topped, pyramidal head, when growing free, but only a small, narrow crown in the forest. *Bark*, dark brown, or gray tinged with red, separated into broad, flat ridges by deep, narrow fissures. *Twigs*, gray, stout, two opposite each other; *buds*, rusty-brown and felty, leaf-scars semicircular.

Habitat. — Rich or moist wood.

6. The Paper Birch, or Canoe Birch. *Betula papyrifera.*

Tree fifty to seventy feet high, forming, while young, a narrow, pyramidal head of short, slender, spreading branches, with long, drooping branchlets; old trees often supporting a round-topped, airy head of pendulous branches.

Bark on young stems and large limbs creamy white, easily splitting into thin, paper-like layers, of which the inner ones are bright orange; much broken and dark near the base of old trees.

Branchlets dull red the first winter, gradually changing to a darker orange-brown, with increased lustre, for about five years, then turning white.

Male flower catkins form in summer and remain on the branchlets until spring.

Habitat. — Rich woods, shores of lakes, banks of streams. From the bark of this tree the Indians built their canoes.

Read : How to build a canoe, in The Young Voyagers, by Reid.

7. The Ironwood or Hop-Hornbeam. *Ostrya Virginica.*

This is a small tree, usually not more than twenty to thirty feet high. In its summer foliage it is sometimes mistaken for a young elm. In its winter condition it is easily distinguished from the elms by its brown, cylindrical flower catkins, which appear in fall, but do not open until spring.

The bark is about one-fourth of an inch thick and broken into narrow, light brown scales, slightly tinged with red on the surface. The buds, fully one-fourth of an inch long, seem rather large for the slender branchlets. Both buds and branchlets are of a dull brown color and quite smooth, but scarcely lustrous.

The wood is very hard and heavy, and often shows a twisted grain. Let the boys relate their experience in cutting and splitting it.

8. The Wild Hazelnut. *Corylus Americana.*

Shrub, three to ten feet high; bark brownish-gray; forming copses along edges of woods, along fences, and wherever trees and shrubs grow; staminate catkins exposed during winter. Compare the catkins of birches, alders, hazel, and ironwood.

9. The White Elm. *Ulmus Americana.*

This is one of our most beautiful native trees. It is often from one hundred to one hundred and twenty feet high, and its large, spreading branches, with slender, drooping branchlets, often form a crown a hundred feet or more across. Sometimes you find in grainfields magnificent, solitary elms with a beautiful, symmetrical crown seventy or eighty feet above the ground. Do you think that the farmer pruned the tree of its lower branches and thus compelled it to form that grand crown which you can see at a distance of many miles?

The bark of the white elm is ashy gray, and irregularly divided by deep fissures into broad ridges.

The branchlets of the last season are dark brown and minutely hairy. The second year they assume a grayish-brown color and are smooth but not lustrous.

The buds are dark brown and smooth. Those which contain flowers increase in size very early in the season.

The Slippery Elm, *Ulmus fulva,* has *downy,* rust-brown buds; the Corky Elm, *Ulmus racemosa,* also has downy buds, but its young branches are covered with a *rough, corky* bark.

10. The Butternut, or Oilnut. *Juglans cinerea.*

A fine tree from fifty to seventy-five feet high, generally dividing fifteen to twenty feet from the ground into numerous stout limbs, which spread horizontally and form a broad, symmetrical, round-topped head. Bark, strong-scented, gray or brown, deeply divided into broad, flat ridges. Young trees and branches smooth and light gray. Branchlets stout, generally dusky-green the first year; slightly hairy, gradually becoming gray; leaf-scars remaining very conspicuous for several years; pith of the twigs light to dark brown, consisting of transversely arranged plates; buds terminating branches fully half an inch long, the others much smaller.

Habitat and range. — Rich woods; from New England to the mountains of Georgia; west to Minnesota, Kansas, and Arkansas.

11. The Bitternut, or Swamp Hickory. *Carya amara.*

Tree with a tall, straight trunk; fifty to seventy-five feet high; with stout, spreading limbs; and slender, stiff, upright branches; forming a broad head, if not crowded. Bark light brown to light gray, broken into thin scales. Branchlets the first winter, orange-brown, lustrous, with large leaf-scars, ultimately light gray. Terminal buds bright yellow-green and large.

Habitat and range. — Moist soil; from New England to Florida; west to Minnesota, Nebraska, and Texas.

12. The Bur Oak, Mossy-cup Oak. *Quercus macrocarpa.*

About eighty feet high, but often much smaller. Stout branches, some growing nearly at right angles from the stem, forming an open irregular head, when the tree is young, and a massive, round-topped crown in older trees. In the forest the crown is much narrower. *Bark* one to two inches thick, deeply furrowed and broken into irregularly light brown or gray scales on the surface. *The branchlets* are stout and marked with pale lenticels, and are at first covered with a soft felt. When they are from two to four years old, thick, corky wings frequently begin to develop.

13. The Scarlet Oak. *Quercus coccinea.*

A tree sometimes seventy feet high, but usually not averaging more than thirty feet, with comparatively small branches, which generally form a somewhat narrow, open head. The bark of the trunk is almost black outside, but red internally, and is divided by shallow fissures into irregu-

.ar ridges. The branchlets are light red during the first winter, and change later into light or dark brown. Young acorns, about the size of a small pea, can be found on the branchlets in winter. These acorns mature the following summer.

This tree has the most beautiful scarlet foliage in autumn, which changes to brown in winter, many of the leaves remaining on the tree until spring.

14. Hackberry, Sugar Berry. *Celtis occidentalis.*

Tree about twenty-five to forty feet high, with spreading, sometimes pendulous branches, which form a symmetrical, round-topped head. Bark of the trunk light brown or silvery gray, very much broken on the surface, sometimes with irregular, wart-like excrescences. Branchlets slightly zigzag, slender, often two or three close together near the ends of small branches, red-brown and marked with oblong leaf-scars the first winter, gradually becoming dark brown. Buds ovate, pointed, flattened by the pressure of the stem.

15. The Linden, or Basswood. *Tilia Americana.*

Tree about forty feet high. Bark dark brown, much furrowed, the surface broken into small, thin scales; inner bark fibrous and very tough. Branches slender, often pendulous. Twigs at first greenish or reddish brown, but soon turning dark gray. Buds about one-fourth of an inch long, and nearly as thick, rich red-brown or cherry, glossy, slightly mucilaginous.

§ 3. Review and Summary.

We have learned that all of our native trees, with the exception of evergreens, have no leaves in winter. The leaves, as well as the flowers for next season, are contained in a rudimentary state in the buds, which are protected by scales. These scales protect the tiny leaves and flowers against sudden changes in temperature, and keep rain and snow from them. Some buds are naked, others covered with hairs, others with a kind of varnish or resin. Is this resin soluble in water? Find out which trees have two buds opposite each other, and which ones have them arranged alternately. Some trees have a bud terminating each branchlet; others, like the elms, never show truly terminal buds. Study the arrangement of buds on different trees, and verify or disprove the statements made here. Each tree has its peculiar mode of branching. Could you classify trees according to the soil and locality where they most commonly grow? Where do willows and poplars generally grow? Where would you expect to find maples and oaks?

The best time to plant trees is late in fall and early in spring. The more roots can be left on the tree, the more likely it is to grow. The branches should be trimmed, but should not be cut away entirely. Most of our trees can easily be raised from seeds and many will grow from cuttings. You should try all these different ways of raising trees, for that will teach you much which you cannot learn in school.

III

THE WOODS IN BLOOM. MARCH TO MAY

§ **4.** MATERIAL : Winter twigs and flowering twigs of all the trees to be studied. Review briefly what the class learned about the flowers of willows and poplars last spring.

1. *Aspen flowers.*

When I told you some weeks ago that you should look out for the flowers which would soon appear on the trees, some of you thought that most trees had no flowers ; but this morning Fred brought us a bundle of flowering twigs, which he picked from a tree. Can you tell me on which tree they grew ? You are right; they grew on the aspen. As we have studied many flowers, I think we can easily make out the structure of these. Fred is ready to describe his catkins " My catkins are about an inch and a half or two inches long; from a distance they look gray, but this color is due to many little scales which are divided into several lobes, fringed with soft gray hairs. Among the scales I find a large number of little scales, on each of which from six to twelve stamens are inserted, and each of the stamens carries an anther filled with pollen. When I shook the poplar tree, the broken anthers discharged their pollen like a mass of dust." Now let Annie tell us about her catkins: "My catkins are also provided with hairy scales and little discs, but on the discs I find small conical ovaries, of which each carries two small styles." Fred found the two kinds of catkins on different trees, and he asked me if both kinds

Observations. — Continue to observe the birds mentioned previously.

would produce seeds.   We will all watch the tassels for the
next two or three weeks, and I think we can then answer
his question.   I think you will find that the staminate ones
will soon wither and fall off.

The class might now be asked to compare the flowers of
some other poplar with those of the aspen.

FIG. 26.   SOFT MAPLE.
a, fruit; *b*, leaves; *c*, flowers.   All reduced.

Observations. — Note also chipmunks, red and gray squirrels, rabbits,
and other small mammals.

2. *The flowers of the Soft or Silver Maple.*

The boys found some more tree flowers yesterday. Do you know from what tree these twigs came? As you have all recognized them as twigs of the soft maple, let us see if the flowers differ much from poplar flowers.

Susan finds at once two kinds of flowers; both kinds grow in clusters from lateral buds, and they are all attached to thin filaments; but one kind contains a bunch of dark red anthers carried on short stamens, while the other kind has two small wings in each flower. The two wings are styles, and are attached to the ovaries in which the seeds will grow later. We look in vain for a bright-colored corolla in maple flowers, but you can easily find the greenish-yellow calyx below the stamens and also below the ovaries. You remember that we always found the staminate and pistillate flowers of poplars on different trees. On maples the two kinds also grow generally on different trees, but sometimes both kinds occur on the same tree. Who has branches that bear both kinds of flowers? On your way home you must look for bees and other insects on maple flowers, and try to get some flowers of the rock or sugar maple for to-morrow.

3. *The flowers of the White Ash.*

The boys say the sugar maple has no flowers yet, but they found other tree flowers. Do you know these twigs? Minna is right; they are twigs of the white ash. Here again we find two kinds of flowers, and the boys tell me that the staminate and pistillate flowers grew on separate trees. Do you find the minute calyx below the stamens and the ovaries? Who will describe the shape and color of the anthers? What is the color of the pollen? The ovaries are surmounted by long styles, which have each two dark purple stigmas. We called the flowers of poplars catkins; loose flowering branches like these are called panicles. Are you not surprised that the large trees which we have studied bear such small, inconspicuous flowers ?   I wonder if we shall find any trees with large, bright flowers.

FIG. 27. WHITE ASH.
a, twig with leaf and fruit; *b*, staminate flowers; *c*, pistillate flowers. All reduced.

4. *The flowers of the White Elm.*

The flowers of the white elm are *perfect;* that is, each flower contains both stamens and pistils.   They appear before the leaves, from buds larger than the leaf-buds. Several flowers are attached to a short, common stalk by long pedicels. The calyx is green and has from seven to nine lobes, the ovary being provided with two green styles, stamens

FIG. 28. WHITE ELM.

A twig with leaves and fruit, below a twig with flowers. Both reduced.

protruding from the calyx. Watch the little ovaries; they are mature fruits when the leaves have unfolded.

Did the maples, poplars, and ashes have perfect flowers?

5. *The flowers of the Paper Birch.*

The staminate catkins are produced in two or three flow-
ered clusters, which are formed in the fall. They are from
three to four inches long, and in early spring, when they

FIG. 29. PAPER BIRCH.
a, twig with leaves; *b*, flowers; *c*, fruit. *a* and *b* reduced.

open, they shed an abundance of yellow pollen. The pistillate catkins are much smaller and appear in the spring; their scales are light green, the styles bright red. Both kinds are found on the same tree.

FIG. 30. IRONWOOD.

A twig with leaves and fruit, below a staminate catkin. Reduced.

6. *The flowers of the Ironwood.*

The sterile catkins appear in the fall, one to three together at the tip of short branchlets. They are about two inches in length when they open in spring. The loose, fertile catkins appear in spring and terminate the leafy shoots of the season.

7. *The flowers of the Hazelnut.*

The long, slender tassels of the hazel bush are among the first signs of spring. The sterile catkins appear in summer and open in the following March or April, shedding yellow pollen in profusion. Several of the fertile flowers are concealed in one scaly bud; each ovary bears two elongated, bright red stigmas, which are easily found if the bushes are looked over with some care.

The flowers of the hazel come out very early. Let the pupils look for them when the poplar catkins are opening. There are two species of hazel common everywhere from the Atlantic coast to the Dakotas, and from Canada to New Jersey, — the common wild hazel (*Córylus Americana*) and the beaked hazelnut (*Córylus rostrata*). Both form extensive thickets and copses, and in some years yield a rich harvest of sweet nuts for boys and girls as well as for squirrels and chipmunks. The common hazel is a shrub from two to four feet high, the nut is enclosed in two leafy bracts, which are open when the nut is ripe and are not prolonged into a beak. The beaked hazel is a shrub from two to six feet high, and has the nut enclosed in several united bracts, which are prolonged into a slender, tubular beak.

8. *The flowers of the Rock or Sugar Maple.*

The flowers of this maple appear with the leaves, as you see. The pedicels, or stalks of the separate flowers, are produced in bunches of long, hairy threads. Find the calyx, the stamens, anthers, and the ovary with its styles. The sterile and the fertile flowers are in separate clusters on the same or on different trees. The seeds ripen in autumn. Compare the flowers of the soft maple, and if your time permits, also the flowers of the box elder.

9. *The flowers of the Butternut.*

Staminate and pistillate flowers, separate, but on the same tree, appearing when the leaves have attained about half their full growth; staminate flowers in catkins three to

FIG. 31. SUGAR MAPLE.

a, twig with leaves and fruit; *b*, staminate flowers; *c*, pistillate flowers. All reduced.

five inches long; fertile flowers six to eight on a terminal shoot, surrounded by bracts covered with sticky hairs; two bright red, spreading stigmas half an inch long.

FIG. 32. BUTTERNUT.

a, pistillate flowers; b, twig with leaf and fruit; c, staminate flowers. All reduced.

Find out from observations whether the staminate or the pistillate flowers mature first.

10. *The flowers of the Bitternut, or Swamp Hickory.*

Sterile flowers in clustered catkins; fertile flowers two to five on a terminal shoot; stigmas green, maturing before the staminate flowers open.

FIG. 33. BITTERNUT, OR SWAMP HICKORY.

a, staminate flowers; *b*, pistillate flowers; *c*, twig with fruit and leaf. All reduced.

11. *The flowers of the Bur Oak.*

Flowers rather late in May or early in June, when the leaves are about one-third full grown; sterile flowers in little clusters on thread-like catkins; fertile flowers on slender stalks from leaf-axils; fruit maturing the first season.

FIG. 34.   BUR OAK. — *a*, twig with leaves and young acorns; *b*, staminate flowers; *c*, pistillate flowers in the leaf-axils.   All reduced.

Oaks which mature their fruit in one season belong to the group of White Oaks; those on which the acorns remain very small the first summer and do not mature before the second autumn are Black Oaks. Gray's "Manual" describes nineteen species of oaks from the northern United States east of the Rocky Mountains. The same common name is often applied to different species.

12. *The flowers of the Scarlet Oak* resemble closely those of the bur oak; but the pistillate flowers grow in pairs on short pedicels from the leaf-axils, and small acorns can always be found on last season's wood, which shows the tree to be one of the Black Oaks.

FIG. 35. SCARLET OAK. — *a*, twig with leaves and young acorns; *b*, pistillate flowers; *c*, staminate flowers. All reduced.

13. *Hackberry flowers.*

Sterile and fertile flowers on the same tree; stamens and greenish calyx lobes five; ovary provided with two recurved stigmas; flowers appearing in early spring soon after the leaves.

FIG. 36. HACKBERRY.

Twig with fruit and leaves, reduced; to the left, a flower somewhat enlarged.

14. *Linden flowers.*

The stem of the flower cluster is attached to a leafy bract four to five inches long, and of a conspicuous, whitish color. The flowers are much sought by bees, and if you

look carefully you can see minute drops of honey in them.
Do you find calyx, corolla, stamens, and pistils in each
flower?   Are the flowers fragrant?

FIG. 37. LINDEN.
Leaves, young fruit, and flowers. Much reduced.

§ 5. Review and Summary.

In our study about tree flowers we learned some unex-
pected things.

Most trees flower very early in spring, several even before
the leaves appear. Can you name such trees now?   Most

of the flowers are devoid of any conspicuous colors and of
fragrance, and people who are careless observers often fail
to find them, or to recognize the flowers as such. As these
inconspicuous flowers are not adapted to attract insects, of
which, moreover, very few are found at this time, they must
depend on the wind for the scattering and carrying of their
pollen, and that they are excellently adapted to wind-fertilization we must at once acknowledge. The sterile flowers,
containing the pollen, are produced in enormous numbers;
the pollen is fine and is easily shaken from the long hanging catkins or the protruded stamens. If, however, the
flowers did not come until late in the summer, most of the
pollen would fall on the large leaves, and the chances for
its reaching the stigmas of the fertile flowers would be much
reduced. Study the foregoing very carefully, and determine
for yourself by observation how much of it applies to the
different tree flowers which you have studied. Does it apply to the linden flowers at all? Are they inconspicuous
and not fragrant? Are they visited by insects, or are they
wind-fertilized? How much later do they appear than the
flowers of maples, birches, and elms? On what trees and
shrubs did you find most old birds' nests when you studied
the winter conditions of trees? Could you recognize any
of the nests? Why is it so much more difficult to find the
new nests now?

We might begin at once now and study the fruits and
leaves of our trees also, but as many of these trees do not
mature their fruit until next fall, we will leave this work
until then; but we must all carefully watch the growth of
the fruit and leaves. However, the boys tell me that the
poplars, the willows, the soft maple, and the elms have
large fruits now, and I will ask you to bring me these as

Observations. — Preserve some pine flowers in alcohol.

soon as they begin to drop. Some we will keep for next fall, and the others you may plant and watch them grow.

NOTE. — A teacher who can find the time might very profitably devote a month or more to the study of "The Flowers of Early Spring." Some of the following common flowers are suggested for such work, but, for want of space, a description cannot be given here.

1. The Skunk Cabbage. Our earliest flower.
2. The Pasque Flower. *Anemone patens.*
3. The Marsh Marigold, or Cowslip.
4. The Rue Anemone. *Anemonella thalictroides.*
5. The Hepatica, or Liverleaf. *Hepatica triloba.*
6. Some common Violet.
7. The Bellwort. *Uvularia grandiflora.*
8. The Wild Ginger. *Asarum Canadense.*
9. The Jack-in-the-Pulpit.
10. The Trillium, or Wake-robin.

For the identification of common and conspicuous flowers, the teacher might consult: Dana, How to Know the Wild Flowers. The best work for the identification of plants is Britton and Brown, Illustrated Flora, published by Charles Scribner's Sons; three volumes. This work gives accurate figures of all the native flowering plants which are now known from Virginia to Newfoundland and from the Atlantic coast to the one hundred and second meridian.

IV

ANIMAL LIFE IN THE WOODS.   JUNE AND JULY

§ **6**. A morning ramble over the bedewed meadows, through the shaded woods, or along the sunny beach would not be half as enjoyable if the bobolinks were silent, squirrels not scolding, catbirds and thrushes not singing, woodpeckers not hammering; and if the stately blue heron, the wary duck, the swift-winged snipes and gulls had not taught us to watch with expectancy every clump of rushes and every concealed nook and bay.   All who love nature certainly regret that for most regions of our country the days have passed when the farmer boy could watch as many deer slowly drawing out of the thicket as he can now find gray squirrels slyly peeping at him from the spreading boughs.   But if the larger animals have retreated into the unsettled wilderness, we must study the humbler ones that have remained with us so much more carefully.   If the large game animals have retreated before advancing civilization, our common song birds have apparently much increased in numbers and prefer a settled country to the wilderness.   If you would see and hear many of the winged singers, you must not look for them in the depth of a large forest, but in orchards, in copses, groves, and along the edges of woods.   There you will find the slate-colored catbirds; the brown, long-tailed thrasher, the orioles, the rose-breasted grosbeak, and many of the small warblers, vireos, and sparrows, as well as the bright-colored screaming blue-

154

jay. If you have taken your rambles in the woods and fields of afternoons, you must change your habits and start out in the morning at about sunrise or before it, and you will be surprised at the display of gay plumage with which the woods are alive; while your ears will enjoy the music of Nature's orchestra at its best. Return to the same spot in the afternoon, and you will see little of the birds and hear less. We shall now study a few of the birds we so frequently meet on our walks and rambles. If possible, good pictures or mounted birds should be at hand to impress shape and color upon the children's mind; where this material cannot be had, close outdoor observations must suffice, as it is not advisable to bring freshly killed song birds before a class of children. We shall begin our bird studies with the catbird, which is as common in gardens and orchards as in the woods.

NOTE. — *Supply* the time for the spring arrival and the fall departure of birds from your own observations. The time given in the text refers to central Minnesota. Cheap and good colored plates of our common birds and of many foreign birds have been published in the magazine: Birds, Nature Study Publishing Co., Fisher Building, Chicago. These illustrations are perhaps the best colored plates accessible to common schools. A common opera-glass is a great help in the outdoor study of birds.

§ 7. The Catbird.

MATERIAL : As suggested above, disinfected nests of the last season ; insects and fruits on which catbirds feed. Observed outdoors: Nest-building, song, and feeding of catbirds. See Farmers' Bulletin, No. 54, about the food of all our common birds.

Description. — Along the edges of woods, in bushes near roads, in orchards and shrubbery about your homes, you often saw a dark, slate-colored bird, smaller and more slender than the robin and more graceful in his movements.

Observations. — Continue to observe the mammals mentioned previously.

He has the habit of flitting from bush to bush, where you cannot easily observe him, but where he can the better observe you. If you come too near his nest, he utters a long-drawn-out zēáy, zēáy, which has sounded to some people like the low mew of a cat and has given him the name catbird. If you are careful, you can approach him close enough to distinguish his black tail and crown from the dark slate back. If you sit down and observe him perched above you, you can see that he is dark ashy gray below and marked with a distinct rust-red spot under the tail.

The catbirds arrive in the last week of April, and begin to sing and build their nests as soon as the days become warm and spring-like. Early, before sunrise, perched on a sapling or concealed in a thicket, they pour forth their melody, which I shall not attempt to describe, but shall simply ask you to go and listen to it.

The nest is made in a tangle of Virginia creeper or wild grapevine, or in some thick bush seldom more than six or seven feet from the ground. Let us see what the catbird uses for house-building. First a foundation is laid of sticks and twigs, and on this the bulky nest is constructed of straws, leaves, and grass; the inside is lined with hairs, fine roots, and other soft material. Soon four or five deep greenish-blue eggs are laid, and then incubation begins. The young are hatched after about two weeks, and from then on, until they are able to take care of themselves, both of the old birds are very busy to procure food for their hungry little ones and for themselves.

What they eat. — It is by no means easy in most cases to know just what a bird picks up on the field. But if you are sharp and fortunate observers, you may see our friend glean

---

Observations. — Watch the growth of Indian corn, wheat, rye, barley, potatoes.

caterpillars from the shrubs and trees, or he will catch a grasshopper or a cricket on the meadow; weevils, too, and beetles, and thousand legs are also welcome food whenever he can catch them.

It must be acknowledged that catbirds do some fruit-stealing. Like boys and girls, they appreciate all kinds of berries, and when they cannot find an abundance of these growing wild, they will try those in our gardens, where they seem to prefer mulberries to any other fruit. As we are willing to pay men for services rendered us, why should we not cheerfully exchange a few berries for a swarm of noxious insects and a grove full of music and beauty?

After the young have attained the power of flight, cat-birds are less commonly seen, because they retreat into the woods as the wild berries begin to ripen. About the first of October they generally go South from this part of the country.

Let the pupils supply the dates of the arrival and departure from their own observations. How do young catbirds, robins, and blackbirds differ from chicks?

Children must not visit any nest while the birds are brooding; for many birds will forsake their nest if they have been disturbed only a few times. *That no boy should make a collection of birds' eggs is self-evident.* One of these boy collectors does more mischief among birds than a dozen cats. A collection of typical nests, taken after the young birds have left them, is desirable for the school; but no collection of eggs should be made, because that would accomplish little good and do a great deal of harm. Professional ornithologists and public museums need such collections, but boys and girls and common schools do not need them.

Look for the following weeds:

§ 8. The Brown Thrasher.

Material and previous observations will be the same as for the catbird. Most of the following lessons on birds will be given only in outline, the abbreviated language of which must not be used in the class.

OUTLINE

1. *Description.* — Male and female distinctly reddish-
brown above, wings darker; below, yellowish white, with
brown spots on breast and sides; tail very long; bill black,
almost awl-shaped, like that of the catbird and the robin;
bird a little longer than the robin; more shy than catbird
and robin; female paler than male.

2. *Time* of arrival and nest-building.

3. *Nest* made of twigs, pieces of vine, grass, rootlets; in
low bushes on the ground. Use an old nest to illustrate.

4. *Song.* — Describe from observation. if you can. Is it
superior to the catbird's and the robin's?

5. *Food.* — Principally worms, caterpillars, grasshoppers,
crickets, beetles, and bugs; cultivated fruit only about eight
per cent of the whole; a little grain; wild cherries, elder
berries, and other wild fruit. The above list shows the bird
to be decidedly useful, and it should be rigorously protected.
Much of its insect food is gathered in brush along fences,
the favorite hibernating place for injurious insects.

6. *Time of fall migration.*

§ 9. The Baltimore Oriole.

Material and previous observations as suggested under catbird and
brown thrasher.

OUTLINE

1. *Description.* — Considerably smaller than the robin.
Male: Head, throat, upper half of back, and wings, black.
Below, bright orange; also orange spots on wings and tail.
Female: Dark olive above, dull orange below.

Observations. — Wild mustard, pigeon grass, yellow foxtail grass, wild
oats, common milkweed.

2. They arrive in this region about the middle of May, and begin nest-building a week or two later.

3. *Nest.* — A perfect pocket, woven out of milkweed fibres, strings, threads, cotton, and similar material; suspended near the end of a long, swaying branch, from twenty to thirty feet from the ground.   An oriole's nest is one of the wonders of bird architecture; give every child an oppor-

FIG. 38. BALTIMORE ORIOLE.

tunity to examine it carefully.   Can you identify a bird's nest without seeing the bird?   Have you ever watched a pair building their nest?

4. *Song.* — Rendered in words by Mrs. Wright as "Will you? Will you really, really, truly?" Female does not sing.

5. *Food.* — Worms, caterpillars, grasshoppers and other injurious insects, and some fruit. By means of its food, beautiful color, and song, one of our most beneficial birds.

6. The parent birds are very much attached to their home and young, and bravely defend them against chipmunks, squirrels, and predaceous birds.

7. They leave, at St. Paul, about the first of September.

§ 10. The Rose-breasted Grosbeak.

Material and observations as above.

1. Appear about the first week of May; build their nests the latter part of May or the first part of June.

2. *Description.* — Male : Breast a conspicuous rose-carmine, which extends under the wings. Lower side, rump, some tail feathers, and two spots on wings, white. Female :

FIG. 39. ROSE-BREASTED GROSBEAK.

Brownish, yellowish under the wings, no carmine breast, very thick bill.

3. *Nest.* — A perfect circle, composed of bits of vines, small sticks, roots, grasses, fibres, and other material, lined inside with similar but finer material; placed on a small tree or a bush from five to ten feet from the ground.

4. *Song.* — About the finest of our native birds in this region; a delightful, rolling warble, often heard towards evening, but not rare at any time of the day, except about noon.

5. *The stout, cone-shaped bill shows that the bird is a seed-eater,* but it includes a large number and variety of injurious insects in its diet, and is the only bird that eats the potato beetles in any large number. Cases have been observed where a family of grosbeaks ate every beetle that could be found in a small field. The bird is therefore one of the most useful to farmers, and should be protected everywhere. If the writer's observations are trustworthy, the bird has lately increased very much in this part of Minnesota. In the spring of 1897, dozens of them could be seen on one day.

6. They cease to sing about the middle of August, and go South about the middle of September.

§ 11. Our Wee Birds.

Before taking leave of our feathered friends, we must, at least, spend a few hours in the bushes with the little folks of bird-dom. In the garden shrubbery, in the willow thickets lining the prairie stream, and wherever boughs and bushes sway to the summer zephyrs, you will hear, often also, see them, but only when they make no effort to conceal themselves. You may hear the red-eyed vireo sing in the grove all morning, but as soon as you come near the spot whence the sweet, soft music sounded, the song emanates from another part of the grove. Most of these small birds belong to one of the following four families: Sparrows, Finches, Warblers, Vireos. The first are about the size of the house sparrow, their strong bills are cone-shaped, and grayish and brownish tints are their predominating colors; they are really seed-eaters, but feed their young on insects. The finches are represented by the pretty black and yellow American goldfinch, which will be described below. The warblers are among the smallest of our birds; most of them are brightly colored, have a sweet lisping song, but are so

M

elusive in their habits that only a few are generally known. Their slender bills indicate that they live on insects; the Yellow Warbler, often called Wild Canary, is one of the most easily recognized warblers. The vireos resemble the warblers in habits, size, and plumage, but are much their superiors in musical ability. Nests of the four birds described below must be shown, and the birds must have been observed.

§ 12. The Chipping Sparrow. *Spizella socialis.* (*Hairbird, Chippie.*)

Here I have a nest, which is one of the prettiest I found last fall. It was fastened to the branch of an evergreen by means of a few cotton threads; it is made of grass and completely lined with horsehair, by which you can always identify the bird.

Who knows the little birds that built it? They wore a dark chestnut cap and had a gray stripe over the eyes. Their back was dark brown and striped, the under parts were gray, and they had a black bill and light feet. From their call note, chip-chip, they are called Chipping Sparrows. They seem to love the company of men, and often build in bushes and in small and large trees near houses. The young chippies are restless babies; frequently they tumble out of the nest before they can fly, and then the parents feed the fluttering little ones on the ground. Last summer I saw a parent deliberately lead a young bird, which could not yet fly, from an open place to a corner, where some brush, tall grass, and corn offered shelter against the sun and the cats.

§ 13. The American Goldfinch. *Spinus tristis.*

To-day I have brought another pretty nest. You see at once that a small bird built it. I took it late last fall from a young box-elder crutch near the sidewalk; the young birds had not left it before the last week in August. ·Do you know of birds that breed so late? Let me see if any of you can recognize the little architects from my description. One of the birds, the male, wore a tiny black cap; its wings and tail were also black, marked with a little white; the rest of the body is of a clear, beautiful yellow. If you had not already learned something about the scarlet tanager and the rose-breasted grosbeak, you might be surprised that this bright-colored little bird has a mate whose attire consists of a very modest, brownish-olive above and a pale yellow below. How many of you know the bird I described? It is called the American Goldfinch. Do you not think that it has received a very appropriate name?

How many of you have seen the goldfinch late in fall? Although none of you have seen them, they are with us until November and some remain all winter. But in September the male doffs his black cap, his bright colors fade, and soon he looks just like his mate and the young of the season. If you are close observers, you may even then recognize it by its peculiar dipping flight. The birds return to us in March or early April, but the spring moult is not completed before the latter part of May, and before that time you are not likely to observe the bright yellow and the distinct black of their plumage. Have you learned about another bird which shows such a marked change in color?

Now let us study this pretty little nest. How trim and small it is, and still, how strong and how securely placed on its support. The outside they made of grass and moss; but this material was evidently not considered soft and warm enough for the little baby finches, and the parents therefore lined the nest with fine down and wool. Do all birds of the same species build their nests just alike, or are they governed to some extent by circumstances and by the material which they can most easily procure? I wonder where the Chimney Swifts built their nest, before this country was settled by white people.   Now they nest almost exclusively in chimneys; however, occasionally their nests are found in hollow trees; still, they seem to like a chimney very much better.   Can you discover a probable reason for this ?

The song of the goldfinch is a wild, canary-like warble; its call note is ker-chee-chee-chee, whew-é, whew-é.

§ 14. The Yellow Warbler.   *Dendroica œstiva.*

This little bird looks very much like a tame canary; the male is yellow all over, but brighter below than above, while the female is of a darker color.

The nest may be found in the garden, field, swamp, or forest.   It is one of the prettiest structures built by the skilful little feet and bills of birds.   Take a stroll in late May through the brush adjoining fields and meadows; soon you will flush the golden bird of summer; then carefully search the brush and you will find a perfect baby bird's cradle firmly lashed to the twigs in the fork of a shrub, about four or five feet from the ground.

Procure several old nests, and find out for yourself how the nest is constructed and what material the yellow warbler uses.   Do you think that he is influenced to any extent by circumstances and locality ?   The song of this bright bird, as is the case with all warblers, is very feeble; it sounds like, sweet-sweet-sweet-sweet-sweet-sweeter-sweeter.

You may find in the nests of warblers and other small birds two kinds of eggs.   The larger egg, nearly an inch long, whitish, and speckled with brown and various shades of gray, is a cowbird's egg, which you should take out and destroy.   The cowbird is commonly seen in pastures, following the cattle.   The male is of a glistening black, but has a dark brown head and throat; the plumage of the female is of a dull, brownish gray.   Cowbirds build no nests, but lay their eggs in the nests of other and generally smaller birds, whose young then have to starve, because the stronger cowbird does not let them have any of the food brought by their parents. Cowbirds are often called blackbirds by farmers; their call note is, clucksée!

§ 15. The Red-eyed Vireo. *Vireo olivaceus.*

This bird is a little larger than the yellow warbler and the American goldfinch. Unless you come very close to it, you will see none but plain shades of brown and gray on its plumage, — colors which predominate on hundreds of our small birds. You cannot fail to identify it by its song, which is superior to that of other birds of its size. In almost every grove in the city, as well as in the country, its song is heard from early morning until the evening twilight. Wilson Flagg calls him the Preacher. "You see it — you know it. Do you hear me? Do you believe it?" he hears the vireo say in an emphatic staccato. If you remember these words, you will recognize the bird every time you hear it. Its red eyes and the pensile nest will also help you to pick it from its many gray and brown coated neighbors.

The nest closely resembles a small cup. It is always typically pensile from a slender, forked branch, and may be looked for from three to twenty-five feet from the ground. It is made of different bark fibres, bits of paper, pieces of wasps' nests, vegetable down, and similar material. The three to five eggs are white, marked with brown spots on the larger end.

§ 16. Closing Remarks on Birds.

We have had many occasions to observe that most of our common song birds are the best friends of farmers and gardeners, and of all persons who are interested in the preservation of shade trees and shrubbery.

To form an idea how many insects are destroyed by birds, you must watch a pair feeding their young. You will find that for hours caterpillars, grasshoppers, and other insects are brought to the nest at intervals of only a few minutes. From such observations you can calculate approximately the number of grasshoppers, for instance, which a brood of robins would eat in two weeks or during the whole summer. Let several of the children make such observations, and let the class know the results.

Birds, however, are our friends in more than one way. By their song and beautiful plumage, they contribute much to our happiness and cheerfulness. They are inseparably connected with bright flowers and green woods, with cool shade and mellow sunshine. They add beauty and poetry to life, and every child loves them as naturally as he loves the flowers. But if you would attract them to your home, you must plant some trees and shrubs, and must allow some wild vines and bushes to grow. Most birds prefer small woods, copses, and groves to large, dense forests. In large forests birds are numerous only around the edges and near other openings. Since our country became settled, the common song birds have multiplied considerably, because the number of their natural enemies, as foxes, minks, wildcats, etc., has been much reduced, and because thousands of groves in the prairie states now provide almost ideal homes for the birds, where formerly they found no resting-place.

Pupils and teachers who would like to learn more about our birds than space permits us here are referred to Birdcraft, by Mabel O. Wright, which will enable you to recognize about two hundred birds. A Year with the Birds, by Wilson Flagg, forms interesting reading on birds.

A systematic review of the birds thus far studied might properly follow these lessons; but as the material for this review can be procured at any time, the chapter is given later, when material for other lessons cannot be procured.

§ 17. Mammals in and about the Woods.

The trees and the underbrush are not the home of many birds only. If we look closely on the green foliage, in the flowers, in the dead trunks and stumps, and under the half-decayed leaves on the ground, we find so many insects and other small animals that it would require a lifetime to study the life history and the habits of all. But let us now turn our attention to some of the larger animals which still inhabit the woods, together with the birds. Several times, in our rambles, we saw a brisk little animal, which in shape resembled a squirrel, but was smaller than the Red Squirrel. All of you who have seen the little chatterbox know that I refer to the Chipmunk. On sunny days in fall, when the nuts are ripe, and when the leaves begin to change to a warm red or to a bright yellow, I have frequently observed the Red and the Gray Squirrels busily gathering nuts for their winter supply. If you tread softly, and if your ears are keen, you may hear "Brer Rabbit" rustle through the dry leaves, and see his cotton tail disappear in the brush. These animals we shall study now; and we will begin with the smallest of them, with the sprightly and ever-active chipmunk.

§ 18. The Chipmunk. *Tamias striatus.*

MATERIAL : Picture of the chipmunk ; a clean skull ; nuts, acorns, grain, and other seeds, on which it feeds. Life and habits previously observed outdoors.

The chipmunk looks like a small squirrel; and as it is more commonly seen on the ground than on trees, it is also called ground squirrel. Its general color is reddish-brown; along the back run five black and two whitish stripes.

Although the chipmunk has no hands, still it can build

Observations. — Note the flowers of Indian corn, small grain, potatoes, and other cultivated plants.

itself a home, which you will generally find near fences or dead stumps and trunks, or under a pile of brush. But as the chipmunk is too weak to defend himself against his many enemies, he has learned to dig into the ground with his feet and make himself a snug little cave, where he is protected against the cold of winter and also from most of his enemies.

There are generally two entrances to the nest, and you will soon understand why the chipmunk should prefer two doors to one. The nest is very well described by Charles C. Abbott,[1] whose words are here quoted, with some omissions: " The two entrances were at the foot of a large beech tree. A little grass only grew about the tree, and the holes at the surface of the ground were very conspicuous. No attempt at concealment had been made; but this was evidently because there were here, at this time, but few of their many enemies. The right-hand entrance to the nest proper was nine feet distant from the opening at the foot of the tree; but as the passage had a somewhat tortuous course, the tunnel was really about twelve feet long. The nest proper was about twenty inches in length, and perhaps a foot in height. It was lined with fine grass. I had hoped to find more than two passages to the nest and extra cavities or granaries, but there were no traces of them. These supplementary burrowings, or 'storehouses,' I believe now are made quite late in the summer, and are additions to their main burrows, made when it becomes necessary for them to commence storing up their winter supply of food."

Observations.— Flowers of wild mustard, milkweed, and other common field weeds.

[1] This and other quotations from the same author are taken from Charles C. Abbott's Rambles of a Naturalist about Home, by permission of the publishers, D. Appleton & Co., New York.

" Chip " has many enemies, still his life seems to be a happy one. The farmer's cat catches many a chipmunk and gopher, when she has to provide for a family of hungry kittens; the weasel stalks him through the brush, follows him up into the trees, and even pursues him into his burrows. Hawks and owls pounce upon him, while he scampers along the old zigzag fences; and the farmer boys brush him off the fences with cruel clubs, and take their first shooting lessons while gunning for " chip " and his cousins.

But in spite of all enemies, the chipmunks hold their own. They are prolific breeders, five or six young being generally found in a nest; they find their food almost anywhere, *and on account of their small size* they can find homes and hiding-places everywhere.

Because the chipmunk has so many enemies, it has learned, like most wild animals, to be always on the alert. The habits of "chip" have been well observed and described by John Burroughs and Charles C. Abbott. I once more quote from the latter; he writes about a family of eight as follows: " On the 23d of June six young chipmunks made their appearance about the stone wall in the yard. It puzzles me even now when I think of it, to imagine when this family of eight chatterboxes took any rest or kept moderately quiet. Very frequently during that summer I was astir at sunrise, and I always found that these chipmunks were already on the go, and throughout July they appeared to do little but play. They seemed to be playing at what children know as 'tag,' *i.e.* they chased each other to and fro in a wild, madcap fashion, and tried to touch or catch one another, and sometimes to bite one another's tail. Occasionally the tail of some laggard gets a nip and he gives a pitiful squeal which starts them all to chattering. The way in which they scamper along the tapering points of a paling fence is

---

Cabbage butterflies and monarch butterflies, their eggs and larvæ.

simply astonishing; but however mad may be their galloping, let a hawk swoop down or even pass over, and in a moment every one is motionless. If they are on a fence, they simply squat where they are and trust to luck to escape being seen. If on the ground when an enemy is discovered, and not too far from their underground nests, which is not often the case, unless foraging, they will dart to these nests with incredible swiftness, and going, I think, the whole length of the passageway to the nest proper, they turn about immediately and retrace their steps to the entrance, from which they will peer out and, when the danger is over, cautiously reappear and recommence their sports. These creatures, during the summer, play merely for play's sake, and seem to have no more important object in view than amusement. Indeed, so far as I have studied animal life, this indulgence in play, just as children play, and for the same reasons, is common to all animals. Frogs, perhaps, in this respect are the nearest to being old fogies, as I never could detect anything on their part that the most vivid imagination could construe into 'having fun.'"

About the middle of August the chipmunks settle down to business. They now wander about in search of corn, beechnuts, acorns, hickory nuts, and other seeds and grain. With these treasures they fill their cheek pouches until they look like children with the mumps. In this manner they carry large stores of winter supplies into their nests. A part of their provisions is stored in the nest proper; the balance is laid down in the galleries and special storerooms. This work is continued until the first heavy white frosts appear; at which time they largely give up their outdoor life.

For some time, until real cold weather sets in, they live on their stores; but when winter begins in earnest they become torpid, cuddle close together, and go to sleep in their

Observations. — Honeybees and bumblebees on flowers.

cosey nest. This state of torpidity probably continues without interruption until the first warm days of spring, when the snow begins to melt and bare patches of ground appear.

Animals which sleep through the whole or part of the winter without needing any food are said to hibernate. Do you know of other animals that go into hibernation when winter begins ?

As stated above, *the chipmunks reappear in early spring, in March or April.* As they cannot find any food at that early time, they subsist on their stores. The quantity of food stored is often surprising. A gill of wheat and buckwheat, a quart of hazelnuts, a peck of beechnuts, some Indian corn, two quarts of buckwheat, and a small amount of grass seed have been found in one nest and in its galleries.

As the little animals do eat some grain, farmers have generally little love for them ; still they could scarcely work any considerable injury, unless they should appear in very large numbers.

§ 19. The Tree Squirrels.

MATERIAL : Pictures, mounted specimens or skins of the red and the gray squirrel ; a summer nest, some seeds and nuts found in squirrel stores ; clean skulls of squirrels, chipmunks, rabbits, gophers, mice, and other gnawers ; live squirrel in a cage.

The Red Squirrel (*Sciurus Hudsonius*) and the Gray Squirrel (*Sciurus Carolinensis*) are known to every country boy. Besides the two species mentioned, the Fox Squirrel and the Common Flying Squirrel also inhabit the northern United States east of the Rocky Mountains. The fox squirrel is the largest, and its color is generally a rusty gray. But both the gray squirrel and the fox squirrel exhibit much variation in color.

Examine carefully the claws of butterflies and wasps caught on milkweed flowers.

The red squirrel *builds a summer nest* of twigs and leaves, often lining it with moss. The nest is placed in a tree, and to it the squirrel retires at night and in inclement weather. *The young are generally born in a hollow tree,* which is also preferred by the parent for its winter home. In regions where hollow trees are scarce or where the winters are severe, it makes its winter home in burrows, which it tunnels itself.

In late summer when nuts and seeds begin to ripen, the red squirrels, like the chipmunks, begin to think of the future. They gather large quantities of nuts, acorns, and different seeds which they hide in hollow trees, under leaves and brush, and probably also in the ground. *They do not hibernate,* but on mild days visit their storehouses to feed. On very cold days they are not seen, and they no doubt then live on the provisions stored away in their nest. They frequently draw on the farmer's granary for their winter's supply, and will even take up their permanent abode in such places, if the farmer does not vigorously object to the arrangement.

Although the squirrels live principally on a vegetable diet, all of them are fond of an occasional steak or omelette. *They are inveterate destroyers of the eggs and young of our birds, and should therefore not be tolerated around farms and parks.* They also eat insects; in fact, any small animal they can catch and overcome. I have seen the red squirrel open deformed stalks of cottonwood leaves and eat the plant lice living in them. It was at this work for hours and the ground was strewn with the deformed leaves. The plant lice seemed to have been carefully licked out of their nests.

When a squirrel eats, it sits on its hind legs, gracefully turns up its bushy tail, *and uses its fore feet with great dexterity like hands.*

All squirrels are exceedingly well adapted to arboreal life.
Their bodies are light, the legs short and all the toes are
provided with sharp claws, by which they can hold to the
bark of trees and even run along small horizontal branches.
*Their teeth also are well fitted to open and to grind up hard
seeds.* They have two upper and two lower chisel-like
incisors. As you see on this skull, there is a gap behind
the incisors in each jaw, and behind the gap we see a number of molars. With their sharp incisors, they can chisel
open the hardest nuts and gnaw holes through boards.
Their food is ground between the molars by a forward and
backward motion of the lower jaw. Do you think they
could gnaw by moving the jaws sideways as a cow moves
them? And I have another hard question for you. All
boys know that a knife blade becomes narrower by the wearing away of the steel, if you keep it long enough and use it
much. The teeth of horses and also human teeth wear off
with age. While the squirrel is awake, it scarcely passes
an hour without gnawing on some hard substance. How is
it that its teeth do not wear off? *The fact is, they do wear
off quite fast; but they keep on growing just as fast.* If you
ever keep a squirrel for a pet, you must give it nuts and
hard seeds and even pieces of wood, on which it can use its
teeth; for without such exercise these soon become so long
that they hinder the animal in eating. *Squirrels not only
can gnaw, but they must gnaw.*

I am sure we all enjoy the quick graceful motions of the
squirrels and love to have the stillness of the summer woods
broken by their lively scolding. Who would not rather
observe a live squirrel among the shady boughs, than shoot
it and see it lying on the ground dead and bleeding? All
lovers of nature certainly regret that, on account of their
bird and grain eating propensities, they are not desirable in
the immediate neighborhood of our homes. Young squirrels

make pretty pets, but only the flying squirrel becomes
entirely tame; the others are apt to bite their own master,
when they get older.

*The enemies of the tree squirrels are about the same as those
of the chipmunk.*   When a squirrel is pursued by an enemy,
it often jumps boldly from a very high limb, and always
lands safely on another tree or on the ground.   *When sailing through the air, they spread their legs and stretch out their
tail, and sail like a parachute.*   The flying squirrels can do
this best, because their fore and hind limbs are connected
by a fold of skin.

The flesh of all squirrels is edible; but I believe you will
all agree with me, that it is much more interesting to study
a live squirrel in the woods than to eat the dead squirrel.

The teacher should always encourage pupils to make original observations, and should occasionally give them an opportunity to report
on them.   Children never tire of observing plants and animals outdoors.   Let some boy tell how squirrels hide themselves on trees,
when they are hunted or pursued.

§ 20. The Gray Rabbit, Cottontail.   *Lepus sylvaticus.*

MATERIAL: Picture or mounted rabbit.   As the rabbit is a game
animal, a freshly killed and unmangled specimen may be used to good
advantage.   Clean skulls of rabbits and other gnawers for comparison.
Twigs and young trees, from which rabbits have eaten the bark.

In our study of common mammals, we must not omit
"Brer Rabbit."   He is at·home from Massachusetts to
Florida, and west to the prairies of Minnesota.

OUTLINE EOR LESSON

1. Unlike squirrels, rabbits cannot climb trees; but their
long hind legs enable them to run very fast; uphill they
can run better than downhill.

2. Their large incisors show them to be gnawers, or
rodents, like squirrels, rats, and mice; but they have four

incisors in the upper jaw.   They gnaw the bark from shrubs
and young trees.

3. For the greater part of the day they remain in their
form, which is simply a hollow on the ground, and is concealed by tall grass or brush.

4. They live in the woods, in groves, under brush, even
in lumber-yards of large cities.   At night, early in the
morning, and towards evening, they feed on grasses, herbs,
young grain, and vegetables.   In the winter they live on
bark and buds, and often become very injurious to orchards
by killing young fruit trees.   This can be prevented by
covering the trunks for about two feet above the ground
with tar paper.

5. Rabbits do not hibernate.   Their winter homes are
deserted holes of skunks, woodchucks, and other burrowers.
In the Northern States, they probably do make holes **for**
themselves.

6. When pursued, they run swiftly to some cover, which
may be brush, a hole, or a hollow tree; for they have no
means of defending themselves.   When captured or wounded,
they utter a cry resembling that of a child in pain; at no
other time is their voice heard.   How do they compare
with the chipmunk and the squirrel in this respect?

7. In autumn, the summer hair falls, and a thicker covering
grows.   Along the back this coat is gray, varied with black,
and more or less tinged with yellow; below it is white.
The tail is cottony-white, hence the name.   Does the mixture of gray, black, and yellow make Brer Rabbit conspicuous on the ground?   The cottontails do not become
white in winter.

8. Are not easily tamed; always try to escape from
confinement.

9. *Are among the most defenceless of animals.   Their
enemies are legion.*   Men and boys, dogs and domestic cats,

wolves and foxes, wildcats and lynxes, bloodthirsty weasels, hawks and owls, and even large snakes, all seek to take Brer Rabbit's life.

10. *In spite of all, the gray rabbit holds on, and has even increased with the settlement of the country.* How is he protected? His color blends with that of the ground, the dead grass, or the brush. He is small, and hides where most of his enemies cannot follow. He is always on the alert; the long ears perceive the slightest suspicious sound. Rabbits are very prolific. In northern states young are produced three times in one summer, each litter consisting of from five to seven tiny cottontails. Their food is almost everywhere abundant, and although many of the young fall a prey to their enemies, quite a number survive. The settlement of the country has reduced the number of their natural enemies.

Do you not think that a rabbit suddenly appearing before you, and whisking into the brush, is a pretty sight? How lonely the woods would be without birds, chipmunks, squirrels, and rabbits! Would you have the heart to take the mother from young animals, and cruelly starve the little ones to death? I know you would not. But that is precisely what you do, if you hunt in the season when birds and other animals have young. It is not wrong to hunt for sport, and use the flesh or skin of the animal taken. Still, I believe that it is much more pleasant and interesting to ramble through woods and marsh with sharp eyes, a field-glass, and a kodak, than with a murderous shotgun or rifle. But the boy or man who kills anything and everything, without a good reason for it, is a brute. Unless animals are decidedly injurious, they have a right to live. The earth was not made exclusively for man. Let not nature's ruler degrade himself to be nature's hangman!

The White Rabbit, or Northern Hare (*Lepus Americanus*),

occurs from New England to Minnesota. It is a little larger than the cottontail. During the summer its fur is brown, but becomes white in winter.

On our Western prairies occurs the Jack Rabbit, or Prairie Hare (*Lepus campestris*). It is the largest of our American hares. The cottontail rabbit is also a true hare. Jack Rabbits are yellowish-gray in summer and white in winter. Of what benefit may such change of color be to the animal?

Some time ago, when we studied the grasses, we learned what an important place these plants occupied in the economy of nature. We have recently studied several representatives of a very important and widely distributed animal family, the Rodents or Gnawers. The hare belongs to the larger rodents; the largest North American rodent is the Beaver. Rodents are found in all countries and in all zones. Most of them, as you know, are small, like gophers, rats, and mice. To the rodents belong also the different kinds of Gophers, the Muskrat, the Woodchuck, and the Porcupine. Some of these animals we shall study later. In the meantime, observe the rodents as much as you can, and try to find out in what respect they are important in nature's economy. You have learned that most of them are small and rather defenceless; nevertheless they are very numerous as individuals, and are also in species the largest family among mammals. How do they hold their own against hosts of enemies?

§ 21. The Underbrush.

MATERIAL: Twigs, sprays, flowers, and fruit of a number of shrubs, vines, and small trees. Names of species are not very important for this lesson.

All the trees are now in full foliage and protect the ground from the rays of the summer sun. It would not be possible to raise a crop of wheat or potatoes on ground

N

shaded by dense foliage; still there are plants which can not only endure shade, but which will not live where they are not shaded.   If you transplant ferns, the large yellow lady's-slippers, and many other plants from the woods into your garden, they will die, unless you provide shade for them.   Under the giants of the woods, we find many small trees and shrubs, such as wild gooseberries, dogwood, hazel, prickly ash, and others.   Here and there we see vines that have climbed up on large trees and some even encircle thick trunks with firm coils; the latter kind are called twiners. Can you tell how the wild grapevine and the Virginia creeper ascend to the tops of the high trees?   The false bittersweet, which brightens the bushes in the fall with beautiful scarlet fruit, and the wild hops are true twiners.

All these smaller woody plants constitute the underbrush in woods and forests.   *This underbrush adds much beauty and variety to the woods,* for a whole forest showing nothing but large trees might be grand, but would soon tire our eyes.

The underbrush is also of much importance to all the denizens of the woods.   Do you remember where we found the nests of catbirds, thrashers, and warblers?   Where could they hide their nests from their numerous enemies, if vines and plum trees made no dense tangles and hazel and dogwood formed no thick bushes?   When the young birds are hatched, the numerous insects living on the brush furnish just such tender morsels as the delicate little stomachs can digest.   And later in the season, when the young birds can take care of themselves and possibly begin to tire just a little of bugs, grasshoppers, and caterpillars, then the bushes offer bird cherries and chokecherries, wild plums and grapes, raspberries, blackberries, and all kinds of berries.   Even in late fall, after the woods have again prepared for their winter rest, the thick, black bunches of the twining moonseed, the dark blue berries of the Virginia creeper, and the crimson fruit of the false bittersweet will catch your eye.

In some thick brush the young cottontail rabbits have their layer; there the doe also conceals her beautiful, spotted fawns. In the summer time grass, herbs, leaves, and tender shoots furnish abundant food for all; but during the winter the cottontails have to rely largely on the bark of shrubs and small trees, and the deer on buds and twigs.

The birds that depend so much on the underbrush actually plant many of the shrubs. When they eat the fruit of sumachs, cherries, grapevines, and others, they do not digest the hard seeds and these grow, if the birds happen to drop them in favorable places. During the spring and fall migrations, seeds are often dropped hundreds of miles from the parent plant. Do you think that some trees and shrubs depend as much on birds for the dissemination of their seeds as birds depend on plants for food? What purpose may the black, blue, red, and white colors of ripe berries serve? Are black and dark blue conspicuous colors? Would yellow and brown be conspicuous colors for late ripening berries?

Many shrubs and trees, if they are cut off, will produce new shoots from the stump; others even spread by stems, which grow directly from the roots. Stumps of willows, poplars, lindens, and oaks produce a large number of shoots, which grow luxuriantly and often have abnormally large leaves, because a large system of roots provides food for them. These shoots frequently form a thick brush, but trees and shrubs grown from stumps and roots do not become as large and do not live as long as those grown from seed.

As we observe the trees through the summer, we must also notice the smaller woody plants and their flowers and fruits, although we may not know the names of all.

§ 22.   Closing Remarks on " Life in the Woods."

We must now leave the squirrels, the birds, the trees and shrubs, and turn our attention to the fields.   You have learned only a little about the woods and the life there. Many things which are not found in books you can learn by patient observation, for which most of you have much time and opportunity in your long summer vacation.   Our plant and animal life is just as wonderful and just as inter- esting as that of countries far away in the tropics.   Before we take up the life in the fields, I wish to ask you a few questions, the complete answer to which we shall find when we again study the work in the fall.

You have observed that the snow lingered much longer in groves, copses, and woods than on open fields.   What effect may this retarded snow-melting have upon the water supply of our brooks and rivers?   Do you know anything about the destructive floods that occur almost annually in the Ohio and its tributaries?   We have observed that twigs, branches, and even whole trees fall into streams and more or less retard the current.   What effect must this have upon the water in the streams and the lakes which they drain?   Where will the water after a shower drain off quicker, from a bare field or from the woods? Where will the ground soonest become dry and parched? You can answer these questions from your own observa- tions.   Do trees grow in countries where little or no rain falls?   Which of our states are prairie states?   Have you ever thought about what caused the prairies?

Read: Shaler, The Forests of North America in Aspects of the Earth; Wilson Flagg, A Year with the Trees; and Chittenden, or some other author on the Yellowstone Park. The above books are suited to teachers and older pupils.

V

IN THE FIELD. JUNE TO SEPTEMBER

Note. — The teacher should modify and change this introduction to suit the season and local conditions.

§ **23.** By far the largest part of our country, as you have learned from history and geography, was once forest or prairie. Many thousand square miles of the forests have now been cleared, and several large states have been carved out of the vast prairies in the Mississippi basin. Perhaps your fathers and grandfathers can tell you about the big trees they had to cut down and the stumps they had to grub out, or how they broke up the virgin prairie. *As we walk along a country road or ramble about on a farm, we at once perceive that a great difference exists between the plants farmers cultivate and between the wild plants along roadsides and fences in uncultivated corners and on wild meadows.* On our way to and from school we pass waving fields of Wheat, Rye, Oats, and Barley. These four plants look very much alike before they produce ears, and I shall find out how many of you can distinguish them. Can you tell me what wild plants they resemble closely? Among these small grains we noticed the purple flowers of the cockle and the bright yellow flowers of the wild mustard, and a number of other less conspicuous weeds; but all of them are injurious to the grain crop. We also pass fields of corn which is almost large enough to bloom, and patches of potatoes whose white flowers form a pretty contrast with the dark green leaves.

Observations. — Note the ever-increasing number of insects:

As corn and potatoes are planted in rows, weeds can be kept down among them by cultivating and hoeing.

Along the road, where the ground has been broken up, we meet the common road weeds, with several of which we became acquainted last fall. You may mention a few of them. In several places, where the ground has never been broken, we find plants still different. Later in the year goldenrods and wild sunflowers predominate there, all of which belong to the native wild flowers of our country.

On some of the potato plants we found the well-known injurious potato beetle; on the ground, grasshoppers were common everywhere, and crickets made their music all day long, but could not always be found. A few times we observed a gopher sitting upright and looking at us; but as soon as we approached, he disappeared into his hole.

All of us, I am sure, have seen the kingbird. He generally perches on a post or a rail, and from time to time he darts after some insect, catches it on the wing, and returns to his perch. Some of the boys tell me that they hear Bob White, the quail, every morning, and often see him sitting on the fence post. Ernest tells me that he knows where a flock of prairie chickens live, and that several weeks ago he found a nest with twenty eggs in it.

But when reviewing the plants and animals of the field, we must not forget the soil upon which they all depend for food. We had a good opportunity to observe the soil in several places where road work had been done. On top we found everywhere a layer of black soil, called loam. This loam was from one to one and a half feet thick and contained a very large number of roots. Below the loam was a yellowish clay in most places; in a few localities sand and gravel took the place of the clay.

Observations. — Chinch-bugs, potato beetles, grasshoppers, and others.

§ 24. Wheat, Rye, Barley, and Oats.

MATERIAL: Young grain sown in boxes or cans; young bean-plants; bunches of full-grown grain showing roots and flowers; ripe ears and straws of each kind collected the previous years or selected from early matured plants; a handful of kernels from each kind; small quantities of wheat and rye flour; cracked oats and barley would be interesting to the children. The children should have observed, as far as possible, the sowing, harvesting, threshing, marketing, and milling of grain; but these subjects can best be enlarged upon in connection with the geography of the wheat-producing states. Wherever practicable, city children should see fields of grain. The teacher should briefly review here what has been learned about the structure and use of grasses.

We begin to study to-day *the structure and life history of the most important of all plants, — the cereals.*

Lack of space prevents us from saying much about the history of our cereals. Wheat and barley were known to, and cultivated by, the ancient Egyptians and Hebrews two thousand years before Christ; but these peoples were not acquainted with rye and oats, both of which were, however, well known to the Romans about the time of Christ. Buckwheat did not become known in central Europe until the fifteenth century, when it was probably introduced by the Saracens; the French still call it Saracen wheat. Nothing definite is known about the origin and original home of these plants. Like our domestic animals, they must have sprung from wild species, but these wild forms can no longer be made out with any certainty and may indeed be extinct.

The teacher who is interested in this subject is referred to De Candolle's Origin of Cultivated Plants.

Time for sowing. — In states where the winters are not too cold, as in Iowa and Illinois, farmers raise mostly winter wheat and winter rye. The time for sowing these is in fall. In what month is it done in your latitude? The

How does the kingbird catch its food?

leaves of the winter wheat and also the roots remain quite small in autumn, but the plants remain green under the winter's snow. In more northern states, as Minnesota and Dakota, wheat is sown in spring, as soon as the ground is in condition to be worked. What is the usual time for your locality? Barley and oats are always sown in spring.

Now let us examine *the different young seedlings.* I have here those of wheat, rye, oats, and barley; from all of which I have carefully washed away the soil. They all germinated with *one* thin, grass-like germ-leaf, while a small root proceeded downward from each grain.

Now let us take plants *which are several weeks old.* On these we observe that the original stem has branched or tillered profusely near the ground, so that it looks as if many plants grew from one seed. The one small root which we observed in the very young plants has decayed, but its place is taken by a whole bunch of fibrous roots which spread in all directions, but do not penetrate very deep into the soil. How do these fibrous roots differ from the roots of the burdock, the carrot, and the parsnip?

Culm and leaves. — The culms are hollow, but nevertheless strong and also sufficiently flexible. If you look at one carefully, you will notice longitudinal whitish ribs, which are also well shown in a cross-section of the culm. These whitish ribs consist of a tough and somewhat woody substance. But the principal source of strength for the long culms are the knots, or nodes, of which several are found in each culm. Cut one of these nodes and you will find that it is very hard and not hollow. Another source of strength are the leaves, which form a sheath for several inches around the stem above each node. *Thus constructed, the long, thin culms are able to stand erect.* Since many of them grow

---

Observations. — Observe prairie chickens and quails, but be very careful not to disturb their nests.

close together, they protect one another in a strong wind, and although they easily bend, so that a grainfield often reminds us of the rolling waves of a lake, they become kneed only by a severe rain storm and by hail. If grain is laid by a heavy rain, the nodes shorten on the upper side and lengthen and thicken on the lower side. In this way the culms rise again to enjoy their share of the sunlight and the summer breeze. Only in places where for various reasons the growth is too rank, the grain will not rise again, if it once becomes laid.

Ears and flowers. — The ears of wheat, rye, and barley consist of numerous smaller ears which are called spikelets. The spikelets are attached to the common stem in two columns which are opposite each other. In wheat and rye we find one spikelet on each joint of the stem; in barley, three. A wheat spikelet has three to four, a rye spikelet two flowers, and a barley spikelet has only one flower.

For a large green calyx and a bright corolla we look in vain. Their place is taken by a number of chaff-like scales, which are persistent and later enclose the seeds. In rye and barley, some of the scales end in a long awn; these awns are called beards by farmers. Most varieties of wheat are beardless. On those ears which are now in bloom we find in each flower three anthers, suspended from long, thread-like filaments which surround a young ovary, provided with delicate, feather-shaped stigmas. The flowers of the oats are not produced in ears, but in loose, open heads; in other respects the flower does not differ so very much from those just studied.

Now you may tell me how the flowers of grain are fertilized. Recollect what you have learned about wild grasses.

Plants that produce only one germ-leaf, like wheat, rye, and other grasses, are monocotyledonous plants; those which pro-

Observe the cunning, playful ways of the striped gopher.

duce two germ-leaves, like the bean, are called dicotyledonous. Cotyledon means a germ-leaf. What becomes of the cotyledons of the bean? How can you distinguish monocotyl and dicotyl by their leaves? Hold them up against the light and notice the venation.

In our next lesson we shall study what is perhaps the most important American grain, — Indian corn or maize.

§ 25. Indian Corn or Maize.

MATERIAL: Corn just sprouted and corn plants a few weeks old; similar material of small grains and beans for comparison; stalks with roots, tassel, and silk; ear with corn in the milk; different varieties of ripe corn on the ear.

This plant is one of the most recent additions to the list of our cultivated plants. Maize, often simply called corn in this country, was cultivated in Peru, in Mexico, and to some extent by the Indians in the United States, when white men first came in contact with the natives of these several regions. Kernels of maize have been found in ancient Peruvian and Mexican graves, which fact proves that, at the time alluded to above, the cultivation of maize was not only known over a vast territory, but was also very old. The original home of the plant is probably in tropical South America.

The introduction of maize into all mild and semitropical regions has proved a boon to humanity. About the year 1500 it was introduced in Europe, and is now extensively cultivated in all southern Europe and in Asia Minor. For the United States it is the most important of all cereals, much larger quantities of it being raised than of wheat.

OUTLINE EOR LESSON

1. Compare seedlings of small grain with seedlings of corn.

Observations. — What other mammals are found in cultivated fields?

2. Compare the full-grown plants: size, roots, culm, nodes, leaves.

3. Compare the flowers of corn with those of wheat, rye, and barley.

The "tassel" of corn contains the sterile flowers; look for their stamens and anthers. The "silk" is composed of the long stigmas of the fertile flowers. Find the exact spot from which a thread of silk starts. Pollen falling upon the silk causes the young grain in the husk to grow. How is the pollen shaken from the anthers? Are the flowers well adapted to wind-fertilization? What is an important difference between the flowers of corn and the flowers of small grain? Corn is the only grass which bears the sterile and fertile flowers on separate heads.

4. Corn requires more heat than wheat. Does it grow best in warm, sheltered bottoms or on exposed uplands? How far north in your state and in the United States is it successfully cultivated?

5. Use and importance.

6. *Varieties.* — Some are more hardy than others. Use of different varieties: pop-corn, sugar-corn, and others.

7. Compare the dicotyledonous bean with the monocotyledonous grains and grasses. Hold the leaves up against the light and notice the difference in venation. Could you tell monocotyledons by the leaves alone? To which class do our common trees belong?

§ 26. The Potato.

MATERIAL: Whole plant, showing roots; young tubers; the old, partly decayed tuber; leaves and flowers; plants injured by the potato beetle; adults, larvæ, and eggs of the beetle; mature tubers, showing buds; plants or sprouts grown in dark cellars; seeds from the previous year; and seedlings at different stages of growth would also be very instructive.

History of cultivation. — The introduction of the potato has proved a real blessing to the people in northern and in western Europe. Its home is probably in the mountains of Chile. Although it is now such an important food plant, and is cultivated in all temperate countries where white men live, yet for centuries after its discovery the potato was not generally cultivated in Europe. The Spaniards were the first to bring it to Europe, about 1580, and some of the stories connected with its introduction in different countries are too good to be forgotten.

In France the chemist Parmentier did much to make the new plant known; but the peasants were not at all pleased with the gift he desired to present to them. As laws commanding the peasants to try the new plant could not be enforced, Parmentier hit upon the following device: He rented large tracts of land near Paris, and there cultivated the potato on a large scale. He then announced that stealing any of the foreign plants was forbidden on pain of death, and in the daytime the fields were patrolled by watchmen, but in the evening the watchmen withdrew. By these ostentatious measures the curiosity of the peasants about the new plant was aroused to the highest pitch, and as the fields were unprotected at night, they stole every potato that grew on them. Thus the potato was soon introduced everywhere in France, about one hundred years ago.

The Prussian king, Frederick the Great, who was as zealous to promote the welfare of his country as he was to enlarge its territory, did much to make this useful plant known in Prussia, but the following story shows that his efforts were not always promptly appreciated.

In 1745 he sent a wagonload of potatoes to his good city of Kolberg, where up to that time nobody had ever seen a potato. All owners of gardens from the city and its suburbs were summoned by a royal order to appear at the courthouse, where his Majesty intended to confer a special favor upon them. Great excitement prevailed when the tubers were shown at the meeting; nobody listened to the detailed directions, which a worthy alderman read to the crowd; everybody was busy,—breaking, cutting, smelling, and tasting his raw potatoes. Some offered them to their dogs, who, of course, refused to eat them. "The things have neither smell nor taste, and not even the dogs will eat them; what could we do with them?" was the verdict of the crowd. Many believed that the potatoes would grow into trees from which the fruit could be shaken off like apples. Not a few simply threw their potatoes on the garbage heap, others planted them in the most improper way and in the queerest places; and the first potato crop in Kolberg was a total failure. But Frederick's government looked to the execution of its orders. In the summer Kolberg's aldermen went around on a potato inspection tour, and all those found negligent were summarily fined. They paid their fine, and blamed the poor potato for their undeserved hard luck. In the next year Frederick repeated his gift, and sent an expert potato grower with it. Potato culture was now taken up in earnest, but half a century elapsed before potatoes became a somewhat common crop on the fields of Prussia.

Nearly everywhere in Europe the peasants behaved in the same way. According to their reasoning, a black tuber, which dogs and even pigs would not eat, could not be good for men.

At the present time potatoes are grown in all temperate regions, from Hammerfest, the most northern city of Europe, to New Zealand and Australia; only the benighted Chinese have refused to try them, although thousands of Chinamen periodically die of starvation.

OUTLINE FOR STUDY

1. *Description :* —

Stem. — Show how it differs from that of small grain and corn.

Leaves. — Pinnate, leaflets odd in number, dark green, smooth, hairy, or rough.

Flowers. — Several in a bunch; calyx and corolla five-lobed; anthers five, yellow, large, grown together; one pistil. Compare grasses.

Fruit. — A spherical, green berry, not edible, containing many small seeds.

Subterranean stem. — Provided with scales here and there, from whose axils root-like subterranean branches spring. These branches thicken at their ends and form the tubers commonly called potatoes. The fact that the tubers have buds, from which leafy sprouts grow, shows that the tubers are thickened branches and not a true fruit. Potatoes can be grown from seed, but plants and tubers are very small the first year. Gardeners grow new varieties from seed. The children might grow potatoes from seeds. Just cover the seeds with soil, and keep the soil moist.

Roots. — The true roots are thin and thread-like; distinguish them from the underground branches.

2. *Why a very useful plant :* —

a. It will grow on almost any soil.

b. Is easily propagated by tubers, which yield a vigorous plant the first year.

c. The yield of tubers can be increased by heaping earth around the stems. By what means do farmers do that?

d. It has developed many varieties.

e. It furnishes wholesome food for man and beasts, *but is not alone sufficient as food for man,* because it contains little else than starch and water.

f. It is used in the manufacture of starch. Ask a farmer's wife how you can make starch from potatoes. Put a drop of potato juice on dark paper, and examine it with a good lens; you will see many of the starch grains. What becomes of the starch in the tuber which is planted? Examine such a tuber.

3. *Diseases and insect enemies.* — The tubers in the ground and the foliage above the ground are subject to several diseases, caused by microscopic organism. About these diseases consult your state reports. Its most formidable insect enemy, the potato beetle, will form the subject of one of our future lessons. On what could the potato bugs live, before potatoes were planted in this country?

4. *Remarks.* — The potato crop is never a complete failure. Since potatoes have been generally cultivated, no general famine has occurred in western Europe. In former centuries great famines were quite common. In India and in China they are still common, and a few years ago even Russia experienced a serious famine. Why should a failure of crops cause a famine in Russia, when it does not have that result in England, France, or Germany?

§ 27. Concluding Remarks on Cultivated Plants.

MATERIAL: Show as many of the plants mentioned below as you can procure. Also show the products made from them, if practicable.

The cultivated plants which we have just studied by no means complete the list of those grown in this region. In our fields we also find flax, timothy, clover, millet, and perhaps the sugar beet. Among the corn we observe the long vine and large fruit of the pumpkin. One of our farmers who keeps bees has sown several acres of buckwheat, and how busy the bees are in that field! Some farmers cultivate small fields of beans, peas, and onions, and in warm, sheltered places the boys and girls have their melon patches.

Now, if we could only make a journey to the South, how different the field, the meadows, the woods, and the orchards would look there! Instead of waving wheat fields, we should see immense areas of corn, cotton, sugar cane, and rice. Of our well-known forest trees very few are found there. In the orchards grow the orange, the fig, the almond, and luscious raisin grapes; and rushes and grasses ten to forty feet high form the well-known canebrakes of the swamps. Could we extend our journey to the south of Florida, we should be in the very tropics, where the pineapple, the date palm, the stately banana plant, and the lofty cocoanut palm greet us.

Send to the Department of Agriculture for a list of Farmers' Bulletins, and ask the Superintendent of Documents, Union Building, Washington, D. C., to send you regularly the monthly list of publications issued by the Department of Agriculture. These lists are sent free, and they will tell you how to secure many valuable publications. See Farmers' Bulletin: No. 35, Potato Culture; No. 39, Onion Culture; No. 52, The Sugar Beet; No. 20, Washed Soils. See the Zone Map in the Yearbook of 1894. The Yearbooks of the Department of Agriculture can generally be secured through the congressman of your district.

§ 28. A Few Common Weeds.

1. **Wild Mustard.** *Brassica Sinapistrum.*
2. **Yellow Foxtail, Pigeon Grass.** *Setaria glauca.*
3. **Wild Oat.** *Avena fatua.*

MATERIAL: Whole plants of each species in different stages of development, each showing roots; seeds and seedlings, if possible. Show the plants a day or two before the lesson, and let each pupil provide himself with material. The children should have observed the plants for some time. If wild mustard and wild oats are not common in your region, take other weeds more common and consult your state reports about them. Touch briefly upon the weeds which were studied before.

See Bulletin No. 34, Weeds of the Mustard Family, Iowa Experiment Station.

1. The Wild Mustard.

This is one of the most troublesome weeds in some parts of Minnesota and Wisconsin. It has been introduced from

FIG. 40. WILD MUSTARD. *Brassica Sinapistrum.*
Leaves, seed pods, and flowers. About one-half natural size.

Europe. In some Minnesota counties, where flax has been grown extensively or where the farmers have been careless, the fields are infested with it almost beyond hope

o

of eradication. The writer has seen fields which were yellow with mustard flowers, and the owner with his wife and children had to pull mustard the whole summer for five, six years in succession, before the pest was under control.

Description. — It is a coarse, rough, annual plant, resembling in general appearance the garden radish in bloom, but has a more irregular and branching root. The stem and branches, which are sparsely clothed with leaves, terminate in heads of yellow flowers, of which the lower ones open first. The stem in the meanwhile continues to lengthen, forming a long, leafless raceme, with knotted pods towards the base, open flowers towards the summit, and a head of closed flower buds at the apex. The flowers are typical of the mustard family. There are four petals and six stamens, two of which are shorter than the other four. Compare the flowers of radishes, turnips, and cabbage. The seeds are spherical, resembling those of the cabbage, and have a harsh, biting taste.

How disseminated. — Wild mustard is often sown with flaxseed. After it has once appeared in a neighborhood, the seeds are scattered by the wind, by running water, in manure, and by threshing machines.

How best destroyed. — A careful farmer goes over his fields from time to time and pulls up the plants while in bloom, as the yellow flowers are easily seen. Thus he never permits them to get a hold and the annual labor involved is very light. After the land becomes badly infested. it takes heroic pulling. If it can possibly be avoided, no grain containing mustard seeds should be sown. An ounce of prevention is always better than a pound of cure.

2. The Yellow Foxtail, or Pigeon Grass.

This grass is too well known to need any detailed description. It was introduced from Europe, and is now common in all our states. It grows on broken soil in all crops and on stubble. Small grain, if in good condition, keeps it down; in corn and potatoes it has to be kept down by cultivation. Its name refers to the tawny yellow spikes or ears. Two other species of foxtail are also common, both from Europe. The cultivated millet also belongs to the foxtail grasses.

3. Wild Oats.

This weed is most troublesome from Minnesota to Oregon and California. The species here referred to is distinguished from cultivated oats by its usually larger size, *earlier* and irregular ripening. Each grain falls out as soon as it is ripe, and the first and second floret are each provided with a strong, twisted, and bent awn. The skin or glume enclosing the grain is hairy below the middle, usually nearly black at maturity, and harder and tougher than that of cultivated oats. The grain itself is much lighter than the cultivated varieties of white oats. Compare cultivated oats.

It is most troublesome in oat crops, because it crowds out the true oats and reduces the value of the threshed grain. The stiff, twisting awns sometimes cause trouble by irritating the nostrils and mouths of animals.

How to destroy it. — If a field is not badly infested, it may be pulled or cut by hand, before it matures. No oats should be sown on such fields.

For cleaning out badly infested fields, see Farmers' Bulletin, No. 28 U. S. Department of Agriculture.

§ 29. The Common Milkweed, or Silkweed. *Asclepias Cornuti.*

MATERIAL : Plants at different stages of development, some showing the perennial rootstock ; several flowers for each pupil ; several dead butterflies, moths, bees, wasps, and other insects with pollen masses attached to their feet ; pass the insects around when the fertilization is explained ; pods and seeds, collected the previous season ; flowers of carrot, caraway, parsley, parsnip, and dill for comparison ; some young milkweeds, raised from seed.

Every child knows the milkweed, or silkweed. It is called milkweed from the abundance of white milk which trickles from the plan when it is wounded. It is also called silkweed on account of the long, white, silky hairs which are attached to the seeds.

Description. — The stout stem generally rises from a thick, running rootstock, which is buried deep in the ground. It grows from one to two feet high. The oval leaves are rather thick, four to eight inches long, and minutely downy beneath. The flowers have a peculiar dull purplish tinge; they grow in umbels, which means that the stalks of the individual flowers rise at the same place from a common peduncle or stalk. Other well-known umbel-bearing plants are the carrot, parsley, parsnip, caraway, and dill. The flowers of the elder do not form true umbels.

The flowers of the milkweed family have a very complicated structure, which the children need not study in detail. It is important that they understand how the pollen masses are transferred from one flower to another. The teacher is referred to Mueller, " Fertilization of Flowers "; Gray's " Manual "; and Britton and Brown, " Illustrated Flora." The pupils can understand the following about the structure if they have the material before them.

The five lower lobes are, of course, the divisions of the calyx; the five lobes next above are the corolla lobes. The peculiar hood-shaped bodies with a horn in the centre are appendages to the anthers, whose united filaments form the tube which encloses the two ovaries.

In the plants which we have thus far studied, the pollen consisted of very fine dust-like grains. In the milkweeds it forms yellow, waxy masses. On the disc in the centre of each flower, you will notice five minute black bodies. Pull these out with a pin. The small yellow bags which come out with them are the pollen masses. The teacher should see that every pupil finds the different structures.

I will now pass around to you a number of insects which have peculiar little bodies adhering to their feet, and I wonder if any of you will be able to identify these little bodies. Willie says they are the pollen masses of milk-weeds, and that is just what they are. Where did the insects get them? I caught these insects on milkweed flowers. What did they do there? Look into the little hoods; some of them are half full of honey, and we know now what the insects were after. In searching for the honey, they could not help stepping on the little black bodies. They stick to the insects' feet, are pulled out with the adhering pollen masses, and the pollen masses are broken and stripped off on other flowers. *The claws of insects are the only means by which the pollen of milkweed can be transferred from one flower to another. This shows that the milkweeds are highly specialized for insect-fertilization.*

Scientific investigators have found that it is impossible for a milkweed flower to fertilize itself. Why could the wind or the hairs and wings of insects not transfer this pollen? Place some fresh milkweed flowers where house flies and ants can get at them, and observe what happens.

Besides the common milkweed, which grows everywhere in rich fields, several other species are not at all rare. The most conspicuous one is the Butterfly Weed, *Asclepias tuberosa*. Its flowers are of a beautiful bright orange, and the plant deserves to be cultivated. It has no milky juice.

Another common form is the Swamp Milkweed, *Asclepias incarnata*, of wet places, and with bright purple flowers. Some milkweeds have greenish flowers, but all have their flowers constructed on the same plan, by which they can at once be recognized.

What insects have you found on the flowers of milk-weeds? How many of you have ever found caterpillars with black and yellow stripes on milkweeds? They are the larvæ of the large and very common Monarch Butterfly.

The seeds. — Find by actual count how many flowers there are on one plant. You will find that only a few of these are fertilized and develop seeds.

If you are a superficial observer, you will probably say, " Well, now, that intricate contrivance for insect-fertilization is not very successful after all !" But if you will now take the trouble and also count the number of seeds in a pod, and remember that from two to six pods are produced on one plant, you will be compelled to admit that insectfertilization in milkweeds is eminently successful. From four hundred to fifteen hundred seeds seems sufficient for one plant. There are two ovaries in each flower. Why could not every one of them develop into a full-sized seed pod? Of what possible use can the large number of flowers be to the plant if only a few produce seeds? Imagine yourself a bee and looking for milkweed honey.

Dissemination of seeds. — It is not enough that a plant should produce seeds; these seeds must be scattered in some way, for if they all dropped near the parent plant and remained there, they would perish for lack of space. Examine a ripe milkweed seed and see if you can tell what happens when the pods burst. Let the children describe the structure and dissemination of these seeds. Why could they not be heavy like nuts and acorns? How are heavy seeds disseminated?

The common milkweed is sometimes troublesome in fields. If you know the life history of the plant, you should be able to suggest a remedy. Young milkweed plants are recommended by the United States Department of Agriculture as a good pot herb.

As the milkweeds are prolific seed-bearers, and also spread and increase by means of rootstocks which remain alive 'from year to year, they appear well qualified to compete with other plants, and also to hold their own against man. Of what advantage is the double mode of propagation to a plant? How long would it take a rootstock to creep fifty miles under ground? Could it cross a river? Some plants have almost become extinct near our towns and villages. Do you think the milkweeds will soon be amongst them?

Insects seen in Fields

If we found very many birds that are undoubtedly the farmer's friends, we must say that many insects are amongst his worst enemies. We shall now study a few of the common insects which we have observed in fields. We will begin with the prettiest of them, the butterflies.

§ 30. The White Cabbage Butterflies.

MATERIAL: Live butterflies on a spray of weeds in a glass jar; caterpillars at various stages of growth, feeding on cabbage leaves; pieces of cabbage leaves with butterfly eggs on the lower sides. It is not important that the children should distinguish the native species from the imported, nor the males from the females. The children should have observed the insects in all their stages.

There are two very common species of cabbage butterflies; both have white wings marked with black and more or less tinged with yellow. On the first warm days of spring we see them flying about. "They are indeed harbingers of spring, and they delight the eyes of all observers so much more because insects of all kinds have been absent for the last six months. These butterflies, though apparently such frail objects, can stand a great deal of ill treatment at this period. It is not uncommon to find them frozen and apparently dead, yet they recover as soon as the rays of the sun warm them back to life" (Lugger). Later in the season these white butterflies often become very numerous.

You will observe that they use all six feet for walking, while some butterflies use only four. They have four wings, to which the different colors are given by delicate scales which easily rub off. From the head two organs stand out like horns; these are the feelers, and the insects use them to touch objects as we use our fingers. You cannot fail to observe the large eyes, by means of which they generally see a boy before he is quite near enough to catch them. But where is this mouth? If you look closely just below the butterfly's head, you will observe an organ coiled like a watch spring. On uncoiling it with a pin, you will find it possessed of considerable length. This long sucking-tube is the principal part of a butterfly's mouth. If you have not seen how they use it, you must watch them on the flowers. Most butterflies live on the honey of flowers, but they are often seen in great numbers drinking water on moist ground.

Life history. — The cabbage butterflies lay their eggs on the lower side of rutabaga and cabbage leaves, where they can be found on close inspection. They look like small yellowish pyramids about one-sixteenth of an inch long. They are not found in bunches like mosquito eggs. The eggs are also laid on peppergrass, shepherd's purse, and other weeds, but only such plants are selected as the caterpillars will eat.

The caterpillars. — The eggs hatch within a few days, and the first meal of the very small, greenish-yellow caterpillars are the shells of their eggs; those finished, they proceed to the cabbage leaves. The appetite of most healthy children is so good that they can eat from four to six times a day, but the young caterpillar's appetite is so great that it must spend most of its time in eating. As caterpillars dwell literally on their bread, and eat so voraciously, they must grow very fast. Within a few weeks they are full grown, and each one then weighs probably about one hundred times as much as it did when it left the egg. Would the biggest cow weigh as much as one hundred calves? You can find out by observation within about how many days a caterpillar eats an amount of food which weighs as much as itself. If a man ate as much in proportion to his weight as a caterpillar, how many oxen could he eat up in a year?

The chrysalis. — You know, from observation, that what we have said about the caterpillar's appetite is entirely true ; nevertheless, there comes a time when these voracious eaters seem to get sick. They stop eating, and move about restlessly. This is a sign that they are about to change into the pupa, or chrysalid state. They prefer a place where they are sheltered from rain and snow, and for this reason the chrysalides are frequently found under the top boards of fences, where they pass the winter. From these chrysalides the butterflies emerge on the first warm days of spring. In early summer they frequently pass through the chrysalis stage on their food plant. The caterpillars of this butterfly, as those of all others, moult several times before they attain full growth. When a caterpillar is ready to moult, the skin breaks open on the back, and the caterpillar works itself out of it. Many caterpillars eat their cast-off skin. The processes of moulting, of changing into a pupa, the emerging of the butterflies from the chrysalides, should be observed by the children on specimens raised by themselves, and they may then describe the process in full.

Remedies. — The best friends of the farmer and gardener are insect-eating birds and mammals, and those we should protect.

In order to show how many insects are destroyed by birds and shrews, Professor Otto Lugger had the position of five hundred chrysalides marked, on October 1, at St. Anthony, Minn. Only a few insect-eating birds remain in Minnesota during the winter; but these, together with shrews, had reduced the original five hundred chrysalides to two hundred and seventeen, by April 2. About the first of April, the migratory birds return to Minnesota, and on May 1, only forty-three chrysalides were left.

Some very small wasps lay their eggs into the caterpillars, or the chrysalides, and the little maggots eat their host alive.

If you have observed caterpillars at all closely, you must have seen that they bite off pieces of leaves, and do not merely suck the juice; they can therefore be easily poisoned by Paris green and London purple. A little boiled flour or stale milk should be added to the liquid containing the poison. These poisons cannot be used after the cabbage begins to head.

See First Annual Report of Minnesota Entomologist.

§ **31. The Monarch.** *Anosia plexippus* of Comstock's "Manual"; *Danais Archippus* of French's "Butterflies of the Eastern United States."

MATERIAL: A sufficient number of adult insects, chrysalides, caterpillars, and eggs. Observed outdoors: Large swarms of monarchs in late summer and early fall, caterpillars and eggs on milkweeds. Call attention to the pollen masses often attached to the claws of butterflies.

This is perhaps the most common and conspicuous butterfly in the Eastern and Northern states. In August and September they often flock together in large swarms, and thousands of them may then be found clinging to the leaves of one tree. On a Sunday morning, early in September, 1896, the city of St. Paul swarmed with them. They came in the forenoon, and most of them had left the next day It is very probable that they winter in the South, in the adult state, migrate North in spring, and return to the South in autumn.

Outline for Lesson

Adult. — Upper surface of wings tawny red, veins black, black border of wings, containing two rows of white spots · under side paler than the upper.

FIG. 41. MONARCH BUTTERFLY. *Anosia plexippus.*
Somewhat reduced. After Comstock. By permission of the Comstock Publishing Company.

Eggs placed on lower side of milkweed leaves, about one-sixteenth of an inch long; first white, then yellow, finally, dull gray.

The larva, or caterpillar, hatches in about a week, first eats its eggshell, is about one and three-fourths inches long when mature; head yellowish, body marked with transverse stripes of black, yellow, and white; two black horns pointing backward and two pointing forward.

The chrysalis is about an inch long, bright green, dotted with gold, a band of golden dots extending more than half around the body. It will probably be easiest for teachers and pupils to raise the chrysalides from caterpillars, which are very easily found on milkweeds.

The monarchs hibernate in the adult state. In the extreme south they fly all winter. In what state do the cabbage butterflies hibernate?

Let the children raise monarchs, and let them watch and describe the different stages of development. Try to follow up the life history of several caterpillars; feed them with the plants on which you find them. Are caterpillars at all particular about their food plants? See if you can make the monarch caterpillars eat anything else than milkweed leaves.

For more information about butterflies, see French, Butterflies of the Eastern United States. For insects in general, see Comstock, Manual for the Study of Insects. Scudder gives a very interesting account of the life history of the monarch in his little book: Life of a Butterfly

§ 32. The Honeybee.

MATERIAL: Some comb honey. Ask a bee-keeper to place a few workers, drones, and queens for you in small bottles or tubes; procure from him different kinds of brood cells, some larvæ and eggs. If no bee-keeper lives near, catch some working bees on flowers, and use the cells, eggs, and larvæ of the common wasp to illustrate how members of the bee family raise their young. Any boy can find these open wasp cones behind shutters and boards. Throw hot water on them after dark. Previously observed: Bees on flowers and bees swarming. Make a list of flowers honeybees visit; also of those they do not visit.

Who of us does not associate the buzzing of honeybees with the flowers, blue skies, and gentle breezes of those happy summers of boyhood or girlhood, when life was still a beautiful poem to us, and when our hearts were even lighter than the wings of the bees and the butterflies that hummed and gambolled about us? How the little child, scarcely able to walk, will tumble after the first butterfly of spring! How interested he is, if you show him a bee on a flower and tell him that this little creature gathers the honey he likes so well! He will never forget what you told him, and will want to know more. And we older boys still see in our imagination the big hollow tree, the forest, the streams, the hills that we saw, when we went bee-hunting long, long ago.

The honeybee is one of the very few insects which man has, so to speak, domesticated; but when this was done, we do not know. Into this country the honeybees were introduced from Europe, probably not so very long after the

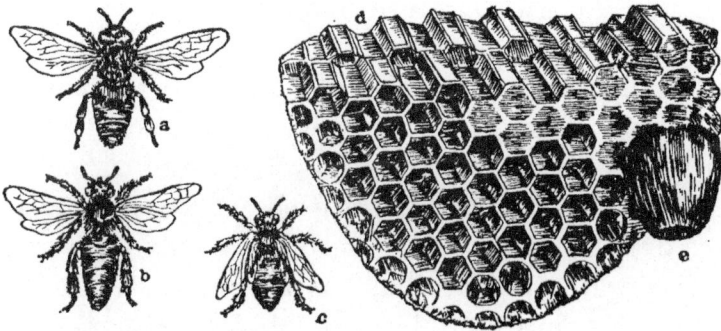

FIG. 42. THE GERMAN, BLACK, OR BROWN BEE.

a, drone; b, queen; c, worker; d, a piece of comb with some open and some closed cells; e, a queen cell. All somewhat reduced.

first permanent settlements. In the large forests, where the old hollow trees offered them so many natural hives, many escaped swarms returned to a wild state. These wild bees slowly advanced westward before the White Man, and about the close of the last century they had reached the Mississippi. It is said that the Indians called them "English flies," and hated them as the advance guard of the White Man himself. They have now spread over the continent, and are found wherever conditions permit their existence.

There are three kinds of bees in each hive. A colony in good condition contains *one queen, from thirty to forty thousand workers, and generally a small number of drones,* perhaps about a hundred.

Function of the workers. — That bees gather the nectar, or honey, from flowers by means of their sucking tongue, you have often observed. This tongue is longer and can be better seen in bumblebees than in the honeybee. If you watch a bumblebee on a cool day, or late in the season, you can best observe its work, because it is not so quick then as on warm days. Bees prefer open flowers, such as linden and buckwheat. In flowers which, like the red clover, form a deep tube, honeybees cannot reach the nectar and must leave it for the longer lips, or tongues, of the bumblebees. Can honeybees reach the nectar in white clover? The nectar is swallowed when gathered, and stored in a honey sac within the bee's body. On coming home, the bee drops it into a cell. But this is only the raw product; it is watery, and often has an unpleasant flavor and odor. By the incessant buzzing of their wings, the bees force currents of air through the hive, and by this draught and the heat of their bodies evaporate the water in the honey down to ten or twelve per cent. Sometimes the bees keep up this buzzing all night. The unpleasant flavors and odors are also more or less driven off, and formic acid, which the bees can secrete by means of glands in their head, is added to the honey as a preservative. After the honey is thus fully ripened, and has been stored in cells near the brood, each cell is sealed with a waxen cap. Ants also secrete formic acid. Press your hand on a number of large ants moving about on their pile, then raise your hand to your nose and you will smell the formic acid.

Gathering of pollen. — We cannot see the honey with which a bee is loaded on its home journey, but we have frequently observed that small yellow or reddish pellets adhere to the legs of bees, giving them the appearance of wearing short but rather wide trousers. These pellets are pollen, which the bees collect on their hairs as they creep over and into flowers. They brush it off as they go along and collect it in a little depression on their hind legs. When the field workers come home, they simply drop the pellets into cells and let the other bees pack them down by kneading them with their mandibles, or outer jaws. This pollen, often called beebread, mixed with honey and water is fed to the brood.

The making of wax. — The little yellow pellets about which we have just spoken must not be mistaken for wax. The bees do not find wax ready in the flowers. The wax is made from honey, and the process takes place in the bodies of bees, the wax appearing in little scales between the segments of the abdomen. It is, so to speak, sweated out in liquid form, but soon hardens on exposure to the air. The bees pluck out the scales with their legs and mould them into any desired shape by means of their mandibles.

Propolis. — This is a brownish substance commonly called "bee glue." The bees gather it from the buds and wounds of trees; they carry it just as they do pollen. It is used to fasten the combs, to make the opening to the hive smaller, to fill up cracks or crevices, and to varnish the whole inside of their home.

When a large brood is in the hive, some bees make frequent trips to streams and pools to procure water, which is needed in mixing the food for the brood.

The structure of combs and cells. — Bees construct four different kinds of cells: cells for breeding workers, cells for breeding drones, cells for breeding queens, and cells for storing honey.

If the teacher has secured these different cells from a

bee-keeper, the children should learn to distinguish them; if that material was not procurable, confine the description to the cells of the honeycomb, which can be bought in the market. That all the different cells are built by the workers needs scarcely to be mentioned.

But the work mentioned thus far does not complete the duties of the workers. *They also defend their hives against robber bees*, which come to steal their supplies. At a time when there are but few flowers, the bees of strong colonies frequently attack those of weak colonies, carry off their stores, and kill thousands of the bees.

As the different duties of worker bees are not equally difficult, the oldest bees perform the more difficult and dangerous duties. The young bees work in the hive for about two weeks; after that time they also fly out for field work. In field work the bees are exposed to winds, birds, and other enemies; thousands of them rip their wings and cannot return home. So great and numerous are those dangers, that the age of a honeybee is not more than about three weeks during the busy season; while wintering, bees may live for eight months.

Duties of the queen. — There is only one queen in each hive, and she is really the mother of the whole colony. Her body, especially the abdomen, or posterior part, is longer than that of a worker. She does not do any work in the hive and never flies out to gather honey, because her weak mouth-parts and weak wings unfit her for that kind of work; nor does she help to defend the hive, although she has a long and powerful sting.

Her only duty is to enable herself, by taking plenty of the most substantial bee food, to lay large numbers of eggs. Under specially favorable conditions, she will lay four thousand eggs in twenty-four hours, and about one hundred thousand in one season. A queen's life is from four to five years, and in that time she can lay about half a million of eggs.

Development of the brood. — According to the cells into which the eggs are placed, and according to the food given the larvæ, the eggs will develop into workers, drones, or queens. The richest food is given to the larvæ in the queen cells. Eggs in the worker cells hatch when three days old; after five days, the white, grub-like larva pupates in its cell. The bees cover the cell with wax and the larva adds a little silk on the inside. Thirteen days later, the pupa has changed into the perfect bee, which bites its way through the cell cover. Queens develop from the egg into the perfect insect in fifteen and one-half days; drones, in twenty-four days.

The drones. — The queen is the only female bee in the hive; the workers are dwarfed females, and, under ordinary conditions, do not lay eggs; the drones are the male bees. They are larger than the workers, but have no sting. When a queen is from five to nine days old, she leaves the hive to meet the drones high in the air. *Mating with the queen and fertilizing her is the only duty of the drones. They do no work in the hive nor in the field to benefit the bee state.* In the fall, when there are no longer any queens to be fertilized, the drones are expelled from the hives or are killed by the workers.

From what we have learned, *it is clear that there exists a division of labor in the bee state; in which respect the beehive may be compared to a community of human beings.*

The swarming of bees may be compared to the founding of new colonies by men.

When the number of bees in a hive has increased so much that they are crowded for room, the workers begin to build drone and queen cells. As soon as one of the queen cells is capped, the old queen becomes very restless, and, with a portion of the workers, leaves the hive, generally before a young queen has emerged. This swarm generally clusters on the branch of some near tree, and is there caught by the bee-keeper. If he fails to be on hand, the swarm will find a new home in a hollow tree or in a rock crevice.

There are several varieties of honeybees. The common wild honeybee is the German, Black, or Brown Bee. Other varieties kept in this country are the Italians, the Cyprians, from the island of Cyprus, and the Carniolans, from the Austrian province of Carniola. Each variety has its peculiar faults and merits.

Conclusion. — We have now learned but very little about the life of the honeybee. Any one who desires to become a practical bee-keeper is referred to Benton, "The Honeybee," United States Department of Agriculture. About the life of wasps and bumblebees, see Comstock's "Manual for the Study of Insects"; but do not forget that the best way to learn about nature is to go to nature's own school.

A Few Injurious Field Insects

The honeybee is a beneficial insect, not only on account of the honey it produces, but also because it fertilizes many flowers, and thereby increases the yield of fruit trees and of many other cultivated plants. Should the two butterflies we have just studied be classified with the beneficial, the injurious, or with those that are neither beneficial nor injurious?

Butterflies and bees are typical insects. Their body consists of three well-marked divisions, — the head, the thorax, and the abdomen. Many insects have wings, which are always attached to the thorax.

In the insects which we are now going to study the three divisions of the body are present, but are not so well marked as in bees and butterflies. Are they well marked in ants ? Do insects have an internal bony skeleton ? Is the outside of their bodies hard or soft ?

§ 33. The Chinch Bug. *Blyssus leucopterus.*

MATERIAL : A number of bugs in different stages of development confined in a bottle with a few blades of grass. By looking carefully, where the grain stands thin, they can generally be found, although they may not have attracted general attention. To show the illustrations and *briefly* to give the life history might be well in districts that are likely to be infested in the future; otherwise omit the lesson where the bugs cannot be found.

The chinch bug is a small insect, only a little over an eighth of an inch long. Its general color is almost black, but the wing covers are white, marked with two black spots and a Y-shaped line. They have the disagreeable odor of bedbugs. Chinch bugs have a sucking mouth, and live on the sap of green grain and different cultivated and wild grasses. They prefer wheat. After the small grain is cut, they migrate to the cornfields.

They winter in the adult state under straw, weeds, brush, and all kinds of rubbish. In spring the reviving bugs move to the nearest food plants, and deposit their eggs on or near the roots. One female can produce over two hundred eggs, and there are at least two broods in a season. Plants which are attacked by many bugs soon wilt and die. It has been estimated that in 1887 the six central grain-producing states lost six hundred million dollars through chinch-bug ravages, and in the same year Minnesota alone lost over six million dollars.

The original home of the chinch bug was probably on the shores of the Atlantic, where it fed on wild grasses; but since the country has been put under cultivation, it has spread enormously, and is now found from New Brunswick

Observations. — Earthworms, their castings and holes, leaves and grasses pulled into the ground by them.

to Florida, and from the Atlantic coast to the Rocky Mountains.

Remedies. — All kinds of rubbish should be burned in the fall, after the bugs have taken up their winter quarters. In cool and wet summers the bugs do not thrive, but are apt to be very numerous under reversed conditions.

For more detailed information, see your State Reports and Bulletin No. 17 of the U. S. Department of Agriculture, by L. O. Howard. For insects injurious in your vicinity, correspond with your state entomologist, and see his reports.

§ 34. The Potato Beetle.   *Doryphora decemlineata.*

MATERIAL : Beetles, larvæ, eggs ; all on potato leaves in a bottle ; pupæ ; defoliated potato stalks.

History. — This insect, often called in books the Colorado Potato Beetle, had its original home near the foothills of the Rocky Mountains in the region of Colorado. Here it fed on a wild plant of the nightshade family, to which also the potato and tomato belong. When the railroad reached that region and settlements sprang up, the beetle discovered the potato plant, found the leaves very much to its taste, and soon acquired the habit of feeding on them. About 1859 it had become a pest in its native home, and began its march eastward across the continent, at first advancing about fifty miles a year, but more rapidly later ; and in 1874 it had reached the Atlantic coast. A few years ago it was even accidentally introduced into southwestern Germany ; but, thanks to the radical and scientific measures adopted against it, its complete extermination was soon accomplished. Why were European insect pests not exterminated in this country ?

Description. — Every farmer knows the beetle, or bug, and you can easily describe the different stages from the material before you.

Observations. — Different soils : loam, sand, clay, humus.

Its life history is briefly as follows : There are three broods in Minnesota, and probably also in the states in the same latitude. The last brood hibernates in the ground ₊in the adult form. Early in spring the beetles fly about, and invade new territory ; and as soon as the first potato plants appear, they deposit their dark yellow eggs on the under side of the leaves near the tips. In about four or five weeks the larvæ are full grown, and descend into the ground for pupation ; and in a short time the adult beetles emerge from the pupæ, and start a second brood.

Remedies. — Kill or poison all the beetles to be found in early spring. A little Paris green or London purple applied early and repeatedly will kill most of the bugs that have wintered. Of course, all farmers should do that. As one female beetle lays about five hundred eggs, you can figure out what its progeny would be for three broods, supposing that one-half of the eggs develop into female beetles. Your figures will show that one lazy and careless farmer can fairly well keep up the stock for a whole county.

§ 35. The Grasshoppers or Locusts.

MATERIAL : A number of large grasshoppers in bottles ; young at different stages of development ; egg masses ; plants injured by grass-hoppers. It is not necessary that the children should distinguish the different species of grasshoppers.

There are so many species of grasshoppers in the United States that no attempt can be made here to distinguish them. For the several kinds of injurious locusts, see your State Reports, and C. V. Riley, " Destructive Locusts," Bulletin No. 25, United States Department of Agriculture.

One of the most common species is the red-legged locust, which measures about one and one-quarter inches from the head to the end of the wings. This species is not migratory,

---

Are the fields in your vicinity well cultivated?

but is occasionally injurious when it appears in large numbers.   Of the migratory locusts, the Rocky Mountain locust
is probably the most destructive.   Several times they have
almost caused a famine in a number of our Western states.
If your region was ever infested, let some old settler tell
you his grasshopper reminiscences, which will greatly interest the children.

The permanent home of these grasshoppers is in the
Rocky Mountains, and from that region they at times mi-

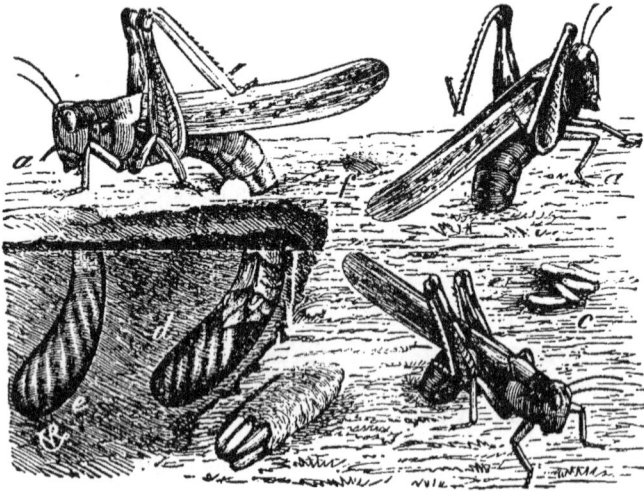

FIG. 43.   ROCKY MOUNTAIN LOCUST.

Showing how the eggs are laid in the ground.   Below, an egg mass taken
out of the ground ; to the right, five detached eggs.   Natural size.
After Riley.

grate in immense swarms as far east as Lake Winnipeg,
Minnesota, Iowa, and Missouri; the frequency of their ravages increasing as one approaches their permanent breeding country.   They do not, however, obtain a permanent
foothold away from their Rocky Mountain home, although
they do deposit eggs and appear in reduced numbers for

Observations. — Does it pay the farmer to keep his fields clean of weeds ?

several years after they have infested a region. The motive which compels them to move is probably scarcity of food in their breeding country. The first swarms eat the more palatable green young grain; if they do not fly off or if other swarms follow them, the insects will eat everything green on the face of the earth.

Life history and habits. — The females dig holes into rather solid soil; closely cropped meadows, fields covered with grass or stubble, and roadsides are the preferred places; and here they can often be found as shown in the illustration. A female probably lays from one hundred to one hundred and fifty eggs in about three different masses. In their native home the eggs laid in summer do not hatch until the following spring, but in warmer regions they often hatch the same summer, and then the young hoppers are often destroyed by frost before they can mature. In Minnesota, the eggs are laid in July and August, and the young come out of the ground the following May.

Habits of the young. — A newly hatched grasshopper looks very much like its parents, but it has no wings. The young Rocky Mountain hoppers are about one-fourth of an inch long. They soon congregate in immense numbers and begin to migrate on foot, generally in a south or southeasterly direction, eating everything green on their way, and if food becomes scarce, the stronger ones eat their weaker brethren. When thus on the march, they do not hesitate to tumble down a cliff or swim a wide river.

The adult insects. — About seven weeks after hatching, these locusts are full grown, and they measure now about one and three-eighths inches from the head to the wing tips. They do not pass through an inactive pupa state, but moult

---

Inform yourself about the work that is being done at the Agricultural Experiment Station of your state and by the United States Department of Agriculture.

about five times. Swarms bred in regions which are only
temporarily infested direct their flight towards the north or
northwest. Swarms which come from the Rocky Mountains, of course move south and southeast. Winged locusts
travel from thirty to thirty-five miles a day on the average,
but with a high wind they may attain a speed of fifty miles
an hour.

Most of the locusts found in the Northern States deposit
their eggs as described for the Rocky Mountain species, and
the habits of the other migratory kinds are very similar to
the species described.

*To the natural enemies of the grasshoppers belong many
parasitic and predaceous insects.* Nearly all birds, as well
as skunks, shrews, gophers, toads snakes, and turtles, feast
on them.

For artificial means of checking locust plagues, see the literature
mentioned above.

Birds seen in the Field

Those of the boys who work in the field in summer and
fall are well acquainted with the Prairie Hen; they recognize their "booming" in the spring; they occasionally find
their eggs and young; but even the girls, who possibly stay
at home more *than they ought to*, have heard Bob White, the
quail, when he was calling on a fence post near the front
door. And the poor city folks who never found time to
leave the big town and see a field of growing wheat, and
who have become so highly educated that they find far
more beauty in shop-window displays than in trees and
flowers, can see thousands of prairie hens and quails on
the market during the shooting season. But there is
another bird which we see so often on our walks through

Observations. — Visit such stations if you have an opportunity.

the fields that we must not omit it here; I refer to the Kingbird.

§ 36. The Kingbird. *Tyrannus tyrannus.*

MATERIAL: As mentioned for birds before. Previously observed: How the bird catches its food, its mode of flight, its nest, etc.

OUTLINE FOR LESSON

This is the bird so commonly seen perching on wires, posts, bushes, and tall weeds along roadsides. From these perches it darts out after a passing insect, captures it on the wing, and immediately returns to its perch.

Description. — Both male and female are black above, with an orange-red streak on the top of the head; the color beneath is grayish-white, and the tail terminates in a white band. The kingbird is considerably smaller than the robin, measuring about eight inches from the tip of the bill to the end of the tail.

The kingbird arrives at St. Paul, Minn., about the first of May; it builds its nest in tall bushes and about buildings, out of twigs, coarse grasses, and weeds, and lines it with fine roots, grasses, and horsehairs. It lays its eggs about the first of June.

Habits. — The kingbird lives exclusively on low-flying insects, catching them, as described above, while they are flying from bush to bush. It is a typical flycatcher like the Phœbe, or Pewee. How do swallows and woodpeckers procure their insects? Compare their habits with those of the flycatcher's. It is a very useful bird, and does not deserve the name of Bee Martin, as it catches only a few drones.

Name. — This bird is a great fighter, attacking and conquering almost any bird that happens to approach it, and is therefore called kingbird.   Let the children tell of their observations on this topic.

It migrates south from here early in September, when the first frosts occur.   Why should it leave so early ?   Do you know of other birds that arrive late and leave early ?   Why do they ?

§ 37. The Prairie Hen, or Prairie Chicken (*Tympanuchus Americanus*), and The Quail, or Bob White.

MATERIAL : Freshly killed birds, if the lesson is given during the open season ; if not, pictures or outdoor observations must suffice. Study these two birds together.

OUTLINE EOR LESSON

Descriptions. — *Prairie Hen:* About the size of a half-grown domestic chicken, but more compact; legs shorter · feathered on the side and in front.   Describe the plumage from the bird before you.

Quail: Only about half as large as the preceding, being about the size of a robin, but body much stouter and tail shorter; legs stronger, fit for running and scratching. Describe the plumage as above.

Habits. — Both nest on the ground; lay from fifteen to twenty eggs. *The young are not fed in the nest, but run about with the mother like young chickens.* In spring and early summer all subsist largely on insects; in fall and winter they feed on seeds, especially on waste grain in the field.   In severe winter weather the prairie hens flock into the woods; the quails find shelter and food about outbuildings, corn bins, and straw stacks.

Both birds have increased much in number since the country became settled, wherever common-sense game laws are enforced. In Minnesota both are more common now than they were in the early days.   The reason for this is that the grainfields afford them a never-failing supply of food, and that their natural enemies, such as foxes, weasels, minks, skunks, hawks, and owls are much reduced in number by man.

Why it is not brutal to hunt these birds in the shooting season. If it is not in itself wrong to eat domestic animals, it cannot be wrong to eat wild animals. But a true sportsman does not kill the old birds at a time when the death of one of them would cause a dozen helpless little ones to starve or be chilled to death, because it would not only be cruel but also a wanton destruction of game; nor does he hunt the young before they can fly well and know how to avoid danger, for to hunt them before that time would be mere butchery, and would also soon exterminate them.

Nature has also, in regard to game and fish, provided for this country so bountifully that there will be plenty of good, healthy recreation for all lovers of the gun and the rod for generations to come, if we only have sense enough not to destroy either without foresight.

Range of the two birds. — The quail is a common bird from the Atlantic coast to the great plains; it prefers districts with some timber; the prairie hen is chiefly found in the prairie states; in the woods its place is taken by the Ruffled Grouse, or Partridge.

Mammals in the Field

In all regions of the earth which are under cultivation by man, large mammals have been exterminated, because many of them would do much injury to crops, if present in great numbers. These large animals cannot easily conceal themselves, and as they are always hunted by men and molested by dogs, they retire to the wilderness. This explains why they are absent from many regions of our country where they would do little or no harm and would find an abundance of food. Only such small animals as mice, rats, gophers, squirrels, rabbits, foxes, weasels, minks, skunks, and others which easily find a hiding-place are common in well-settled districts, and some of them have become very numerous with the increase in their food supply. Of the smaller rodents, mice and gophers are the most common on fields and they furnish a large part of the food for the smaller flesh-eating mammals and for hawks and owls.

§ 38. The Striped Gopher. *Spermophilus tridecemlineatus.*

MATERIAL: Picture; mounted specimen; gopher in cage; freshly killed specimen. Outdoor observations: Cutworms and webworms.

OUTLINE EOR LESSON

Description. — About the size of the chipmunk, but body and especially the tail longer, ears shorter, nails longer, more intended for digging than for climbing. Eight pale yellowish-brown stripes on the back, which alternate with nine yellowish-brown ones; the five uppermost marked with a row of pale spots. Voice, a clear whistle like that of a bird.

Habits. — Lives in holes; hibernates; does not appear until late in spring; ranges northwest from Arkansas and Illinois.

Food. — About forty-five per cent of it consists of insects, especially injurious cutworms; the remainder consists of grain, grass, and green herbs; they do considerable damage by digging up corn before it has sprouted.

Observations at the Iowa Experimental Farm showed that twenty-two striped gophers had eaten from April 19 to August 2 an average number of twenty-six cutworms and webworms daily. Mr. C. P. Gilette, who conducted the

Observations. — Leaves and fruit of the trees, shrubs, and vines you studied in spring and summer.

experiment, says: "Most of the harm they do is strikingly noticeable and is of short duration, while the good is perpetual from spring to fall, but is in no way noticeable and cannot well be estimated."

See Bulletin No. 6, Iowa Agricultural Experimental Station.

§ 39. The Earthworm.

MATERIAL: Earthworms on moist soil in a small dish. Previously observed: Tracks of the worms after a rain; worms creeping about on wet soil, boards, or stones; castings over their holes; leaves and straws drawn into their holes.

You know that birds and mammals have an internal bony skeleton, which gives them a solid frame for the whole body; insects and crabs have no bones, but they have a hard external covering by which the soft, inner parts are protected, and to which their muscles are attached. How much the earthworm differs from all these animals! It has no legs, no bones, no hard carapace like the beetle; its body is simply a bundle of muscles covered by a soft skin.

How can the earthworm move? If you are close observers, you have seen that the whole body consists of a large number of rings, and that the worm can contract and extend itself very much. Along the sides and below you can see, and feel still better, a large number of short bristles, which keep the worm from slipping back as it crawls by extending and contracting itself. Have you ever tried to pull earthworms out of their holes? If you did, you must have found that their strength is considerable. By means of their numerous bristles, and by contracting their muscles, they often escape into their holes when the boys attempt to pull them out.

In spite of a host of enemies, earthworms are abundant wherever a rich soil furnishes them food.

At what time do the various seeds mature?

Moles and shrews burrow for them in the ground; birds, toads, salamanders, and predaceous beetles devour them whenever they appear on the surface. Although earthworms have no organs of hearing, they do perceive vibrations of the ground, which are caused by the walking of men and animals, and these vibrations cause them to retreat into their burrows. They leave their burrows only at night, and then only when the ground is wet, because they cannot move over dry sand or dust. Can you tell why not? Towards morning they return into the ground. They have no eyes, but investigators have proved that a few of the rings near the mouth are able to perceive the light. An earthworm placed in bright sunlight shows by its motions that either the light or the warmth, or both, are exceedingly painful to it. In the morning earthworm tracks can be seen everywhere on moist, bare ground, and on rainy days they can be seen in the daytime creeping on stone and on wooden walks. From which ones of their enemies do their nocturnal habits protect them?

The food of earthworms consists of decaying animal and vegetable matter, such as is abundantly present in all rich field and garden soils. Why do boys not look for earthworms in sterile, sandy soils? The earthworm has no distinct head, but you will find the mouth on the anterior end of the body. They often devour large masses of dirt and extract their food from it, and if there is not sufficient food in the soil, they will pull leaves, straws, and other objects into their burrows and eat them when they begin to decay. Have you seen leaves and straws "planted" by earthworms?

We have learned that the earthworm, although it has none of the sense organs of higher animals, is nevertheless enabled very often to escape from its enemies. *But how can a being of such low organization cope with adverse conditions*

Observations. — By what means are they scattered?

of temperature and moisture? Although it must always have moisture, it cannot live long in water and in very wet soil; therefore earthworms are found near the surface and on higher grounds in a rainy season. In a dry season, however, they retreat to lower grounds and burrow deeper into the soil; and if the drought is prolonged and the lower soil too compact, they coil themselves into little balls, and in this quiescent state await the showers, which enable them to renew their activity. In late fall they go down below the frost line, coil themselves up, and pass the winter in a torpid state. In this latitude they reappear in March or April, according to the weather.

The eggs, from which complete worms hatch, are laid in cocoons in the ground.

In former lessons we learned what important factors insects, birds, trees, and grasses are in the great household of nature. The great English naturalist, Darwin, has shown that even this humble worm performs important work in the vast workshop of nature. He found that the soil thrown up in their castings or pellets varied from seven to eighteen tons on an acre in one year. *It is plain that by the activity of these worms the soil is made finer and rendered more porous, therefore its fertility is increased.* In some places burrowing mammals transfer a very considerable amount of soil to the surface. The writer has observed islands in Minnesota lakes on which every foot of soil had been turned over by woodchucks, skunks, gophers, and other small mammals. Rank weeds soon cover such excavated soil, showing by their luxurious growth that the little folks in fur have put the ground in good condition for plant growth. Have you ever observed how much soil digger wasps and ants, especially the latter, will bring up? From early spring until late autumn you can find hundreds of small ant holes on almost

Insect injuries on leaves, flowers, fruit, and twigs.

any acre of dry ground; every shower washes the excavated soil away and closes the holes, but the next sunny noon finds our pigmy miners diligently at work reconstructing their Lilliputian galleries. When we remember that worms, insects, and mammals had cultivated the soil for thousands of years before man began to use rude ploughs and hoes, it is plain that near the surface there exists probably little soil which these humble creatures have not turned over. *They have done their share to make this earth a place fit for human habitation and human activity.* Should we now begrudge them their homes, where they work and are happy in their own ways, when they do not injure us; should any boy or man be so brutal as to wish to shoot, kill, or crush everything which flies, walks, or creeps?

To our shame it must be acknowledged that there are not only such boys, but also such men still among us.

§ 40. Field and Garden Soil.

MATERIAL: About a quart of black soil mixed with gravel and pebbles; pieces of crumbling and decaying stones; a half-gallon glass jar; a large flower pot. Observed outdoors: Black topsoil; sandy and clayey subsoil; humus from woods or swamps.

It is known to everybody that animals live on the substances they take into their stomachs, and that indigestible matter is passed off with the excrements. It has been a more difficult matter to find out just how plants feed; but careful experiments have shown *that most of the substance which makes the woody tissue of plants is absorbed from the air by the leaves.* If you put a piece of wood into a hot oven until it becomes charred, you will see that wood contains a large part of coal. Green leaves absorb carbon dioxide gas from the air, and in this carbon dioxide gas the coal is contained. If a plant is kept stripped of all its leaves, it will die of starvation. What insects frequently kill plants in this way ? When plants or wood are put into the fire, the coal burns and the ashes remain behind. The substances which form the ashes and the large amount of water present in all living plants are taken from the soil by the roots. *Water and soil are as necessary for plant life as air.*

In order to examine this dark soil in which the plants grow, we put a few handfuls of it into water and shake the whole in this glass jar. The pebbles and gravel at once sink to the bottom when we set the jar down, but the finer soil is mixed with the water and gives it a muddy appearance. After a while the finer soil also settles, and if we let the jar stand long enough the water becomes again clear, but on the surface float partly decayed plant remnants, which are derived either from the plants that grew on the soil or from manure. We can see through the glass that the pebbles and the coarse gravel are at the bottom, and that the finest mud is on top. Now we will put this soil into a flower pot, pour water on it, and thoroughly stir the whole. After the water is nearly clear I shall pour it off carefully, take the cork out of the hole in the bottom of the pot, and then I shall place the flower pot out in the sun, so that the water may drain out and the soil be thoroughly dried. You may observe the drying soil, and to-morrow or the next day we shall study it again.

§ 41. Field and Garden Soil (*Continuation*).

The soil which we placed out of doors is now so dry that it has cracked and shrivelled. If we moisten small pieces of it and then rub them between our fingers, we find that it contains some clay, which makes it sticky, and we also feel and see small grains of sand. But we wish to know what gives the dark color to this soil, and Fred will tell us how he found it out: "I placed a handful of dark soil on a large,

Q

old spoon, and then put the spoon on a strong coal fire and left it there for about half an hour; at the end of that time the black color of the soil had disappeared. As I have often observed that decayed logs and leaves in the woods form a dark mould, I concluded that the dark color in the surface soil is due to decayed and finely divided vegetable matter, and that the dark color of the soil disappeared because this vegetable matter had been burned."

Fred's conclusion is correct, and I wish more of you would perform the same experiment. We find, then, *that the principal constituents of agricultural soil are sand, clay, and vegetable mould or humus.* Besides these the soil contains small quantities of several other substances, such as potash, soda, lime, common salt, and iron. If you bleach out a little ashes, you can taste the potash and soda in the lye, because they are easily soluble. Only such minerals as dissolve in water can be absorbed by plants.

The best soil for plant growth, and therefore for farming and gardening, is a black loam. Loam is a mixture of clay, sand, and humus. Very sandy soil is too porous, and lets the rain water pass through too quickly, and on that account crops are likely to fail on it unless frequent rains occur. Clay is, in this respect, the very opposite of sand. It is sticky, quite impervious to water, and therefore holds the surface water too long in a rainy season, but becomes baked and hard in a dry season. Black loam, however, has just the right porosity for plant roots and for drainage, and still holds a sufficient amount of water and air. On account of its dark color it absorbs the rays of the sun and makes a much warmer soil than clay. *For these reasons it is the best soil for a large majority of plants.*

The crumbling stones which I have here, and the fact that even hard building stones wear and weather if exposed to the atmosphere, to rains, heat, and frost, is evidence that nature is all the time forming more soil. If the wild plants decay where they grew, the amount of humus is also increased. Why do we often find much humus in the woods and very little on the prairie?

From cultivated fields, however, the plants grown are partly or entirely removed, and can therefore not furnish food for those of the next season; and it has been found, as thinking people would expect, that such lands become poorer from year to year, unless some means are taken to restore the lost plant foods to the ground. Therefore the farmer who does not want his farm to become run down restores the plant food to it in the shape of barnyard manure or as artificial fertilizers, and he does not try to raise the same kind of grain on the same field for years in succession. If, for instance, wheat is grown year after year, nearly all of the wheat-producing materials are withdrawn from the soil, and the last crops will fall far short of the first ones.

§ 42. Stock-raising and Agriculture. Neglected or Undeveloped Opportunities in American Farming.

As every successful farmer must fertilize his lands from time to time, he is compelled to keep a good stock of cattle and other animals on hand, because for the large majority of farmers barnyard or stable manure is the best and cheapest fertilizer. Cattle, moreover, enable the farmer to use for pasture or hay land any low meadows which are too wet for cultivation. Bluffs, broken or very stony soil may make good pasture, although the land is not fit for cultivation. Besides the advantages mentioned, a farmer who has a good stock of cattle, hogs, or sheep is not dependent on the price of grain alone for an income. During the last five or six years of low grain prices American farmers have learned to appreciate the wisdom of the mixed farming.

New sources of profit are also being developed. Cream-

eries are now established all over the country, and experiments are being made in the manufacture of beet sugar.
As far as soil and climate are concerned, we can certainly
produce all the sugar we need for home consumption, and
a great deal more. Experiments must determine whether
it is cheaper to produce it at home or import it from abroad.

In many instances American farming is still very waste-
ful, as compared with the best methods followed in Europe.
In Europe the fibre of flax is the most valuable part of that
crop, while the seed is merely a valuable by-product.   In
this country we use the seed only, and burn the valuable
fibre or let it rot. Along roads and in waste places hemp
grows luxuriantly. It has no doubt been introduced with
birdseed, and is now disseminated by wild birds.   Most
people do not even know that this plant produces the seed
for which they pay about ten cents a pound as canary food,
and that it produces an excellent fibre for twine and textile
fabrics. Our wild birds appreciate the seed and also the
fibre. Let the children twist small strings out of flax and
hemp fibre. Why waste the fibre of flax and hemp, and
then pay a big price for imported linen and for manilla
and sisal fibre ?

See the following Farmers' Bulletins: No. 21, Barnyard Manure ;
No. 27, Flax for Seed and Fibre ; No. 40, Farm Drainage ; No. 43,
Sewage Disposal on the Farm ; No. 44, Commercial Fertilizers ; No.
46, Irrigation in Humid Climates.

§ 43. Influence of Agriculture upon Man.

If we wish to inquire into the influence agriculture has
had upon man, we only need to compare the life of white
men with that of our North American Indians. The latter
knew of no "Sweet Home"; they had no fatherland. The
different tribes roamed over their large hunting-grounds,
following the migrations or abundance of the game on which
they depended. They had not even progressed to the stage

of nomadic peoples, probably because game was so abundant that they were not hard enough pressed by hunger to domesticate wild animals and thus procure a constant supply of meat. It is, however, likely that there were still other causes which tended to keep them in the savage state; still we must not omit to state here that some of the tribes cultivated small patches of corn.

When the first settlements of whites were made on the coast of the United States, there were probably not more than five hundred thousand Indians living in a country which now supports sixty-five million souls, and which could easily support twice as many. This shows that by agriculture a much larger population can be supported on the same area than can be supported by the chase.

Hunting tribes and nomadic tribes could never form large states as long as they remained hunters and nomads. History teaches us that all large states have been, and are to-day, based on an agricultural population. With agricultural peoples, the incessant wars waged by savages as well as by nomadic tribes ceased, and the needs of the different seasons taught men orderly and regular habits of work and to look into the future and provide for it. All savages, like our Indians, are very much averse to regular and long-continued work; they gorge themselves when they have plenty of food, but are half-starved when food is scarce. Among an agricultural people industry and commerce spring up naturally, because some apply themselves to making tools, agricultural implements, clothing, etc., for the others, and there we have the beginning of different trades and of factories. As not every person produced everything he needed, a number of persons exchanged their products, and this was the beginning of commerce.

Agriculture enabled large numbers of men to live close together and made progress towards a higher civilization possible.

It has also been found that agriculture has a great effect upon the climate of a country. At the time of Cæsar, Holland, Belgium, and Germany were covered with interminable swamps and forests and had a very cold and cloudy climate. To-day these countries have been transformed into sunny gardens and fields, where all the grains and fruits of the north temperate zone flourish. Old chroniclers and old settlers tell us that in this country the winters were more severe formerly than they are now and that snow and rain were more abundant. Although the different reports of such changes must be accepted with caution, they are no doubt true in general. Can you see why agriculture should make a country drier and warmer? Is such a change beneficial in every respect? How can some of its attendant evils be avoided?

America furnishes us a very interesting chapter in the history of civilization and in the history of mankind. About two hundred and fifty years ago small bands of white people from England, Ireland, Scotland, Holland, Germany, and Sweden settled on our eastern coast and hewed small clearings into the dark, boundless forests. They had brought with them their domesticated animals and cultivated plants. Many weeds, now common, and many insect pests, and also such pests as domestic mice and rats, were accidentally introduced. To-day the white man and the animals and plants which, purposely or accidentally, he brought over with him are in possession of the continent; a continent which is among the most productive, but which nevertheless had never developed a high, general civilization.

Reference Books

The teacher and mature pupils will find the following books very interesting and very instructive : —

Tarr.   Physical Geography. The Macmillan Company. 1897. $1.40.
Shaler.   Nature and Man in America. Scribner, New York. 1891. $1.50.
Guyot.   The Earth and Man. Scribner, New York. $1.75.
Shaler.   Story of our Continent.

VI

THE WOODS IN THEIR SUMMER FOLIAGE.
SEPTEMBER

§ 44. The Leaves of our Trees and Shrubs.

MATERIAL: A walk into the woods, or observations on trees and
shrubs in parks and along streets, should precede this lesson. Leaves
or leafy twigs of all plants to be mentioned ; press and keep leaves of
all. Preserve also a large variety of leaves injured by insects in various
ways.

Most teachers probably have no time to study the leaves more in
detail. The few new terms used should be explained as they occur,
and should be thoroughly understood by the children. Let them
describe the leaves as they see them.

All our trees have had their fully developed foliage for
several months; some are still producing young leaves, but
most of them have only full-sized leaves. You must have
observed that it is much more difficult now to see the
branches and twigs distinctly than it was in March and
April. The birds also know that the leaves protect their
nests from intruders, and they do not build them until the
leaves are out. It is easy, in early spring, to find the old
nests, but not so easy to find the new ones in May and June.

Although the leaves we collected show a great variety of
forms, we can nevertheless arrange them for comparison
in a few groups.

1. The leaves of elms, of the hazel, the dogwoods, the
chokecherry, and the ironwood are more or less oval in

Observations. — Notice the beginning of autumn colors on foliage, and
watch its increase.

shape; have only a short stalk; their margins are serrate, or toothed, but show no large lobes. Which of these are smooth, and which are rough or hairy?

2. The leaves of most of our common oaks are deeply lobed, and have only a short stalk, or petiole. The leaves of most of the black oaks terminate their lobes in a fine, needle-shaped point, while the white oaks show rounded lobes. Are oak leaves smooth, glossy, or downy? Do they look alike on both sides? Do elms or oaks present the more densely massed foliage?

3. Here we have leaves that grew on the sugar maple, the silver maple, and the wild grape vine, respectively. How does the length of the leaves compare with the length of the petiole? How can you distinguish the three kinds from one another? How do these leaves differ from oak leaves, and from the leaves we placed in the first group?

4. The leaves of the quaking asp, the cottonwood, and the birches have a long, slender petiole, and are moved by the slightest breeze. Have you ever heard the lisping, whispering sounds of poplar groves on a summer day? What effect has a strong wind upon poplar leaves? Why should one of the poplars be called the *quaking* asp? Break some poplar petioles by pulling. Why is it necessary that the petioles should be strong? How does the width of these leaves compare with their length? Are the veins more prominent than in maples and oaks?

5. Willow leaves are much longer than wide. Describe their margins. The leaves of the linden are quite broad, and distinctly heart-shaped. How does their venation compare with that in poplar leaves?

All the leaves we have examined thus far had each a separate petiole, which issued directly from the woody twig;

Observations. — Where is the most beautiful autumn grove you know of?

such leaves are called simple leaves. In the next two groups we have compound leaves. Each one of these has a number of leaflets attached to a common petiole.

6. The beautiful large leaves of the asp, the butternut, and the hickory are of this type. The arrangement of the leaflets reminds one of a pretty feather, and therefore such leaves are described as pinnate, or feather-shaped. How can you distinguish these three kinds of leaves by their touch and their smell? Do peas, roses, and prickly ash have compound or simple leaves? Which pinnate leaves have an even number, which an odd number of leaflets? Do you know the Kentucky Coffee Tree? It has the most compound leaves of all American trees. Its flowers resemble the pea flower; and the large, stone-hard seeds are borne in pods that look like very big pea pods. The tree grows in rich woods, especially in river bottoms, but it is not very common.

7. Other types of compound leaves are the palmate, or hand-shaped leaves of the Virginia creeper and the trifoliolate leaves of clover. What does trifoliolate mean? Can you mention other plants that have compound leaves? Are they pinnate, palmate, or trifoliolate? Can you tell, from the position of the buds, if the leaves will stand opposite, or alternate?

§ 45. Insect Injuries to Foliage.

MATERIAL: Galls on oak leaves, rosettes of willows, leaves folded by caterpillars and plant lice, tents of caterpillars, deformed petioles of cottonwood leaves, and other material. Names are of little importance in this lesson; the object is to show the children how largely and variously foliage is injured by insects.

Not only the leaves of small plants are eaten by insects, but sometimes large trees and even whole groves are en-

---

Find places where ferns, horsetails, mosses, lichens, mushrooms, and puff-balls grow.

tirely defoliated by them. Grasshoppers, caterpillars, the larvæ of sawflies and beetles, are all very voracious and become exceedingly injurious when they appear in large numbers.

But even such small insects as plant lice, of which many species are known, may cause much injury. Although they can suck only the sap of plants, they cause the leaves to crumple, to form wart-like growth, or to become abnormal in other ways. Such injured leaves, even if they remain green, are of little or no use to the plant. Folded leaves generally provide both shelter and food for the insects. Some of the injurious insects, such as the red spiders on house plants, are so small that a cursory observer may fail to detect them.

Some small flies deposit their eggs in the tissue of leaves. With the developing insect a gall, or apple, develops, which furnishes the young larvæ food and shelter. Each gallfly selects certain parts of a certain plant and makes galls that are different from the galls of any other species. The galls of gallflies are closed and the larvæ transform within them or eat their way out and transform in the ground. Galls made by mites and plant lice have an opening. It is thought that some poison secreted by the adult insects or by the hatched larvæ causes the abnormal growth which forms the galls.

Trees whose foliage is much injured by insects bear little or no fruit, because their food is used up in producing new leaves. In forest trees the injury consists in a reduced production of wood, and defoliated trees along streets and in parks look bad and give no shade.

See Comstock's Manual; Saunders, Insects Injurious to Fruits; Harris, Insects Injurious to Vegetation; your State Reports. Farmers' Bulletin No. 19, Important Insecticides.

§ 46. The Fruit of Trees, Shrubs, and Vines.

MATERIAL: Fruit and fruit-bearing twigs of all species to be studied ; twigs with small fruit to be added to the school herbarium ; large seeds to be kept in boxes. Dissemination of seeds is important. From the larger seeds remove the coverings and try to find the germ-leaves, the first leaflets, and the radicle, or rootlet. Plant the seeds and observe the young trees.

1. We observed last spring that the little pods on the fertile catkins of willows, of the quaking asp, and of the cottonwood soon opened and shed a large amount of fine white cotton.   Closer inspection showed a tiny seed attached to each tuft of cotton, and *by means of these tufts the seeds often sail many miles and are dropped everywhere.* If they happen to fall on bare, moist ground, they will grow at once. Later, in July and August, I observed thousands of young willows, asps, and cottonwoods growing on the moist sand banks near lakes and rivers.   Where a tract of forest has been burned, a poplar thicket will almost invariably succeed the burned trees.   Later other trees slowly return, and in course of time kill out the poplars or take the place of those which die of old age, as the poplars are short-lived; but it probably takes from thirty to fifty years before this change is well under way.   Why do such small and light seeds as those of willows and poplars have a better chance of finding a suitable place early in spring than later in the season? How are bare sand banks and mud flats near rivers made?

2. The seeds of elms and birches are little nutlets provided with a winged margin, by means of which they are carried considerable distances from the trees.   When were the elm seeds ripe? When were the birch seeds shaken out of the catkins? The seeds of ashes and maples are also provided with wings, and if you drop some from a second or third story window, you can observe how they sail. As these trees grow often near streams, their seeds frequently drop

into the water, are carried away by the current, and finally dropped on sand banks and mud flats, where they can grow. Maple seeds are dropped as soon as they are ripe, but on some species of ash the seeds remain through the winter.

3. Fruits which consist of hard seeds covered by a soft juicy flesh are called berries. Of this type are the wild grapes, the fruit of the Virginia creeper, the blueberry, the gooseberry, the currants, and the fruit of the hackberry tree. You can easily explain why such fruits as cherries and plums are called stone fruits. In the cultivated and in the wild apples the flesh is really an enlargement of the calyx. You will see this if you split an apple lengthwise. *Berries, stone fruits, and the small wild apples depend on birds for dissemination.* Before they are ripe, they are green like the leaves, and very sour; but when they are ready to be planted, they invite the feathered gardeners by a more or less sweet taste and make themselves conspicuous by a black, blue, red, and occasionally by a white color. Why would yellow not be a good color to attract the birds? The birds swallow the small fruits entire; the fleshy parts are digested, but the hard seeds pass through the birds uninjured. As not only winter and summer residents feed on such fruits, but also the flocks of migrating birds, seeds are frequently dropped a hundred miles, or more, from the place where they grew. How could birds plant wild plums?

4. Such large and heavy seeds as acorns and nuts drop right under the tree, and would remain there and decay if gophers, mice, chipmunks, and squirrels did not covet them. These animals often drop them accidentally, hide them under leaves or in their burrows for their winter food, and while engaged in this work they leave some in places where they can grow. The writer once shot a gray squirrel and found that it had a hickory nut still in its mouth when he reached home with it. As hawks frequently catch such

animals, they, no doubt, often plant seeds, which their prey happened to be carrying.

Many birds and mammals depend on trees and shrubs for their food and for shelter.   Do they in turn render any important service to the plants which feed and shelter them ?   What becomes of the young trees and shrubs that happen to grow right under the old ones ?   Have you ever observed the flowers that grow on burnt-over timber land ?

VII

THE WOODS IN AUTUMN. SEPTEMBER
AND OCTOBER

§ 47. Trees in their Autumn Foliage.

MATERIAL : Leaves and twigs of various trees and shrubs with
autumn foliage ; bouquets of leaves from the scarlet oak, Virginia
creeper, sugar maple, sumach, ash, and others may be made very
ornamental in the schoolroom.   If possible, teacher and pupil should
take a walk to the woods and enjoy the beauty of the autumn foliage.
Let the children describe an autumn landscape, which they have observed.   Add the most conspicuously colored leaves to your school
herbarium.   After this chapter has been studied, different pupils
might write up a complete life history and description of different
trees, shrubs, or vines.   If they can illustrate these with simple draw-
ings or paintings, they will be all the more interesting.   In most cases
the teacher should give carefully prepared outlines to the pupils, who
should also have access to the school herbarium.

The working out of the suggested lesson is left to the teacher.

§ 48. Some Plants that have No Flowers.

In early spring we went to the woods for such flowers as
Jack-in-the-pulpit, wild ginger, hepatica, rue anemone, and
bloodroot.   In autumn we find none but very inconspicuous
flowers in places where the above-named flowers with a
number of others told us that the long, dreary winter
had come to an end.   The explanation of this absence of
flowers is plain enough.   These places are so densely shaded
through the whole summer that only shade-loving plants
will flourish there.   But here again we see that nature has
filled every nook and cranny with an abundance of life

Observations. — Study the foliage, fruit, and mode of branching of the
pine, the Norway spruce, or some other evergreen.

238

whenever moisture is not altogether absent. Even light, although necessary for all green plants, is not absolutely needed for mushrooms and those animals which have become adapted to living in caves or in the absolutely dark abyss of the ocean. It is true that in the dense shade of thick woods, farmers cannot raise potatoes, wheat, or corn, nor would we look in such places for sunflowers and goldenrods, for all these plants love the bright sunlight. If, however, we wish to find the delicate fronds of the maidenhair fern or the pigmies among land plants, the mosses or the gray and ever old-looking lichens, then the shady woods are the very places to go to.

§ 49. Ferns and Horsetails.

MATERIAL : As many different kinds of ferns as you can find. With a stout knife or with a hatchet dig up the creeping rootstock of the species you find. Names of species are of little importance to your pupils. Fruiting and sterile horsetails.

Among the most common and best-known ferns are the Brake, or Bracken (*Pteris aquilina*), the Maidenhair Fern (*Adiantum pedatum*), and Clayton's Fern (*Osmunda Claytonia*).

The last mentioned is very common in moist places. In the vicinity of St. Paul it grows abundantly around the numerous ponds in the oak woods. On the fertile leaves some of the leaflets are entirely converted into spore cases. These spore cases contain a green dust, the spores. When the spores are ripe and fall out of the spherical cases, scatter a little of the dust on moist earth in a small flower pot, then cover the soil with an inverted drinking glass and keep it moist, but not wet. Very soon the earth, which should be somewhat firmly packed, will appear more or less green on account of a large number of minute heart-shaped plants

Visit a sawmill and a lumber-yard.

which grow flat on the ground.   This heart-shaped plantlet, called a prothallus, is the first state of a fern.   From some

FIG. 44.   CLAYTON'S FERN.   *Osmunda Claytonia.*

A part of a fertile frond, much reduced ; to the right, some spore cases, slightly enlarged.

of the prothalli, small leaf-like fronds are developed, which will, however, not attain their full size until they are several years old. When they are mature, they again produce the

FIG. 45. BRAKE, OR BRACKEN. *Pteris aquilina.*
A part of the frond, much reduced; to the left a small piece of the frond seen from below, about natural size.

masses of green spores. For various reasons your prothalli are not very likely to develop leafy fronds, but they may, if you keep them long enough and if they are not too

R

crowded.   It is very difficult to find these fern prothalli outdoors, but they do frequently occur near ferns in green-
houses.   The above gives you a part of the life history of
every fern.   The life history of non-flowering plants is ex-

FIG. 46.   MAIDENHAIR FERN.   *Adiantum pedatum.*

A part of the frond in the centre, much reduced; to the left and to the
right pieces of the frond seen from below, the former natural size, the
latter enlarged.

ceedingly interesting, but it cannot be thoroughly studied
without a compound microscope.

Another fern common in and around rather sunny woods
is the Brake.   It is our largest fern, and frequently attains
a height of four feet.   The long stalk bears a large tripar-

tite leaf, and each of the parts is doubly pinnate. The brake has brown spores and spore cases, which are found under the reflexed margins of the leaves.

Among the prettiest of our ferns is the Maidenhair, which is only found in damp, shady woods. The dark, chestnut-brown, or nearly black, stem is sometimes about ten to twelve inches long, but often much shorter. Its two branches divide into many polished thread-like branchlets, which have suggested the pretty comparison expressed in the name of this plant. The shining branchlets are studded with two rows of the most delicate leaflets, under whose reflexed margins the spore cases will be found in little patches. The rootstock, which is about as slender as the leafstalk, creeps along under ground; it is provided with a great number of chaffy scales and has roots along its whole length. Compare the rootstocks of Clayton's fern and of the brake with this rootstock.

Ferns furnish very good material for drawing and for school herbariums; the live plants, as well as carefully pressed fronds, are very ornamental.

§ **50.** Another class of flowerless plants which may be mentioned here are the horsetails. Many of them resemble a miniature pine tree in their mode of branching; others have no branches, but all consist of many nodes, or joints, which can be easily pulled apart. Some species produce fruiting stems very early in spring. These stems are only a few inches tall and *not green,* but of a brownish-white color. They carry a head, whose form is suggestive of a very small pine cone. These heads carry the spores, which, when shaken from a mature head, often cling together in masses by means of fine ribbons, with which they are provided. The species just described produce green, sterile stems later in the season. Other horsetails produce sterile and fruiting stems, which resemble each other.   These kinds produce the dark, spore-bearing heads in early summer.   The spores develop in a manner similar to those of the ferns.

You will find horsetails common on moist meadows and in shallow water; some also grow in rather dry soil, into which they penetrate to a great depth.

All ferns and horsetails in the temperate zones are small plants and are of little importance in nature; but in a former period of the earth's history, when the coal now mined in the coal beds was deposited, vast marshy forests, as large as our present prairies, extended over parts of this continent and Europe.   These forests consisted largely of plants which in structure closely resembled our present ferns and horsetails, but were of gigantic tree-like size.

However strange and beautiful these forests and their foliage must have been, they lacked nearly all of the cheerful change and manifold life of modern woods.   No autumn ever lavished its glowing colors on them; for many centuries they stood in the same monotonous green; no flowers and butterflies enlivened mossy knolls; no birds ever perched and sang on the grand fronds.

§ 51. Mosses and Lichens.

MATERIAL: The species illustrated, or any species that can be found ; lichens gathered on the ground, on stones, and on the bark of trees ; dried tufts of mosses and patches of lichens should be moistened in presence of the children, to show how quickly moisture revives the shrivelled plants, and how much moisture the mosses can absorb ; pieces of wet and of dried peat.

To give the life history of any moss in detail would not be within the scope of this book.   Suffice it to say that the little capsules, which may be found on mosses at almost any time, are the spore cases, and contain green or brown spores.   If the spores are shed on moist ground, they develop a growth of green threads, and from these green threads the leafy stems are again produced. The light spores are shed in dry weather, are wafted about in all directions by the winds, and germinate in all moist and shady places. The teacher who has a good hand lens might try to make out some of the structural beauty of moss capsules and leaves.

The Bare-soil Moss (*Funaria hygrometrica* Sibth.) may be found in spring and summer on moist clayey soil which has been recently laid bare. Look for it on that side of street and railway cuts which is exposed to the north.

The large Hair Moss (*Polytrichum commune* L.) can be found in patches near swamps and in damp woods.

Among the most important moss in the economy of nature is the common Peat Moss (*Sphagnum cymbifolium* Ehrh.). It covers many square miles of swamps, both in this country as well as in Europe.

Fig. 47. *Funaria hygrometrica.* *a*, whole plant, slightly enlarged; *b*, reduced; *c*, a capsule with cap still adhering, slightly enlarged.

Besides the green mosses which you find on moist soil, and on decaying wood and bark everywhere, you must have noticed on the soil, on rocks, on the bark of trees, and on wood, a grayish growth, which sometimes appears in closely adhering patches, sometimes presents a hairy or leaf-like form, the latter often looking like bits of old crumpled paper. These are Lichens. They are generally gray in color, but you will find on them yellow, dark, or red spots and cups. These spots and cups contain the spore sacs of the lichen. The

Iceland moss, which you can buy in every drug store, is a lichen, as is also the reindeer moss, about which you have probably heard.

Scholars who have given many years to the study of mosses and lichens have found hundreds of species of each group; but the principal question with us is *the importance of these plants in the economy of nature.*

They are the first plants which appear on bare rocks, where no life could exist before them.

Soon the lower parts of mosses die, but the dead rootlets and stems furnish a better soil for the green tufts on top. This thin turf holds the rain water for some time, and also catches dust and sand which the winds drop on it. Thus a thin layer of soil is formed, in which larger plants can find a footing, and in which insects and worms can burrow.

Of great importance for man are the peat mosses. For hundreds and thou-

FIG. 48.

To the left, three plants of *Polytrichum commune*, slightly reduced; to the right, a plant of *Sphagnum cymbifolium*, slightly reduced.

sands of years these have covered the extensive peat swamps in our Northern States and in Europe, this year's plants growing on the dead generation of last year. Together with the partially decayed remains of grasses and other swamp plants, they have formed layers of peat from a few feet to forty feet thick.

In Europe this peat is cut or baked into bricks, and much used for fuel; and the time is not far distant when this treasure of nature will be more extensively used in our country.

Mosses, as we have seen, are good absorbents of moisture. They prevent the too rapid drainage and drying of the soil, and thus exert a great influence upon the water supply of streams and upon the life of higher plants. Seeds of wild flowers, shrubs, and trees are frequently embedded in moss, which retains the moisture long enough to cause the seeds to germinate, while on bare soil the delicate leaflets and rootlets would, in most cases, dry up and die.

Mosses and lichens are also important for animal life. In the woods you will find worms, insects, and insect larvæ under the moist green turf; but in Labrador and in the far north of British America, where trees and grasses do not grow, mosses and lichens cover the ground as far as the eye can reach.

As the great grass plains formerly were the pasture of vast herds of buffaloes, and now feed millions of sheep and cattle, thus the great moss and lichen plains support vast herds of caribou and musk oxen. In the high north of Europe the Lapps herd their reindeer on the large moss and lichen meadows. The teacher might tell the pupils about the life of the Lapps, and about the habits of the American reindeer, or caribou.

§ 52. Fungi.

MATERIAL: Different kinds of mushrooms, or toadstools; puff-balls; some hard pore fungi growing on old stumps. Also show some of the thread-like or felt-like growth, which may be found in or on the ground or in dead wood, where these fungi grow. See Farmers' Bulletin No. 53, How to Grow Mushrooms.

All of you, children, have found mushrooms; and some have told me that you could find many more after a rain in the fall. Have you ever tried to find the roots and seeds when you found large specimens, in places where you saw

FIG. 49. A COMMON MUSHROOM.
About one-half natural size.

none the day before? Where did they come from? How did they grow?

Non-flowering plants which have no green tissue are called fungi. People commonly call the large edible fungi mushrooms, while they call the poisonous kinds toadstools. This is, however, not a good classification; and it would be better to speak of edible and poisonous mushrooms.

What a rich mushroom flora we found on our trip to the damp woods. Some had heads larger than big sandwiches, while others were tiny creatures growing on dead leaves, and their heads were scarcely as large as one half of a pea.

In color they vied with the flowers, showing nearly all shades from white and red to black.

Now examine the lower side of the umbrella-shaped heads. We find that some have many ray-like gills running from the stem in the centre to the margin; others show a layer of small tubes instead of the gills. About an hour ago I cut off a mushroom head, close to the gills, and placed the head on a piece of white paper. Now I find a layer of dust-like spores on the paper, which are arranged like the gills. If I had not told you what I was going to show you, you would probably have taken the figure on the paper for a pencil drawing. These spores are the seeds of the mushrooms. They germinate in the ground or in decaying wood, and form a tissue of thousands of fine threads. At the time of fruiting, many of these threads grow together, and produce the well-known heads above ground. The threads, which are called *mycelium*, absorb food from decaying matter, because plants without leaf-green can-

FIG. 50. A PUFFBALL. Reduced.

not absorb carbon dioxide from the air. Have the mushrooms any true roots? The puffballs have no gills, but their interior is filled with a mass of spores. The pore fungi grow mostly on stumps and trees in semicircular or irregular masses. Their tissue is more or less woody; their mycelium can be found in the wood on which they grow. The mycelium of fungi remains alive in the substratum from year to year, but the heads of most kinds appear only at a certain season.

If we are careful observers of nature, our attention must be attracted by the quick and absolute disappearance of fungi, as well as by their sudden appearance.

Cattle, sheep, deer, squirrels, mice, and snails eat them.

Many flies deposit their eggs on old specimens; a fact which explains why these are always full of maggots. In decay, which is very rapid, they emit an odor like that of putrid meat.

Many of our common mushrooms are edible, and it is to be regretted that they are so little used. We should, of course, not eat any mushrooms unless we know that they are harmless, because several species are very poisonous. In districts where mushrooms are common, the teacher should make herself and pupils acquainted with the more common edible species. For this purpose, see Peck, "Mushrooms."

We have learned that the fungi quickly convert decaying wood and leaves into wholesome food for man and animals. Their own tissue decays rapidly, and thus enriches the soil, and furnishes food for other plants.

§ 53. About a Few Other Fungi.

MATERIAL: Wheat rust on leaves; smut of corn; black and blue moulds on bread. With the necessary explanations this lesson might be given after the Grains.

Every farmer boy knows that in wet seasons, wheat and other small grain is apt to suffer much from rust. The brown dust which you can shake from the grain forms the spores of a very small plant, which grows in the tissue of green grain and can be seen as brown spots. There are many different kinds of rusts and they affect many different plants. The black masses of smut are also spores of microscopic plants. Plants that grow in or on the live tissue of other plants are called parasitic plants. Rust, smut, blight, on lilac and willows are parasites in the plant world. Do you know of any parasites in the animal world? Have you ever seen any parasitic plants that bore flowers? The dodders, which form tangles of yellowish or brown threads on wild sunflowers, wild touch-me-not, and willows are such parasitic flowering plants. Look for them along the banks of streams and in other moist and somewhat shaded places.

Besides the fungi mentioned above, we must not omit the various kinds of moulds which grow on bread and other eatables. Dust some of the black or blue spores of mould on a piece of moist bread and keep the bread under a water glass. Within twenty-four hours you can see the white mycelium of the mould on the bread.

For the life history of the fungi mentioned here see Bessey, Botany · Briefer Course. Henry Holt and Company, New York.

VIII

EVERGREENS, OR PINES AND PINE FORESTS.
NOVEMBER

§ 54. The White Pine. *Pinus Strobus.*

MATERIAL: Young shoots with staminate and pistillate flowers in sixty per cent alcohol or pressed ; cones at different stages of development; leafy twigs ; bark of stem and wood. Outdoor observations. The white pine has flowers in June. Almost every teacher can procure material of this lesson from Christmas trees.

We have learned about a number of trees that shed their leaves in fall, and have also learned a little about the forests and groves they form. To-day we shall study a tree which looks very different from an oak, maple, ash, or poplar. These twigs, flowers, and cones grew on the white pine, a tree which has perhaps furnished lumber for more homes in our Northern States than any other tree. In northern Minnesota, in Wisconsin and Michigan, and in Canada this tree often forms large and dense forests, in which the trees attain a height of one hundred and seventy-five feet and a diameter of ten feet. Tell me how many times as high as our schoolhouse such an old giant tree would be, and how many of them could stand on a space as large as this floor.

After a similar introduction the teacher should bring out the following points by comparing the white pine with an oak, a poplar, or any other deciduous tree : —

·1. Regular mode of branching, which approximately shows the age of young trees.

Observations. — Are there still birds in woods and fields?

2. The shape, color, and arrangement of the foliage.

3. The soft, easy-splitting, and resinous wood.

4. Striking differences between evergreens and deciduous trees.

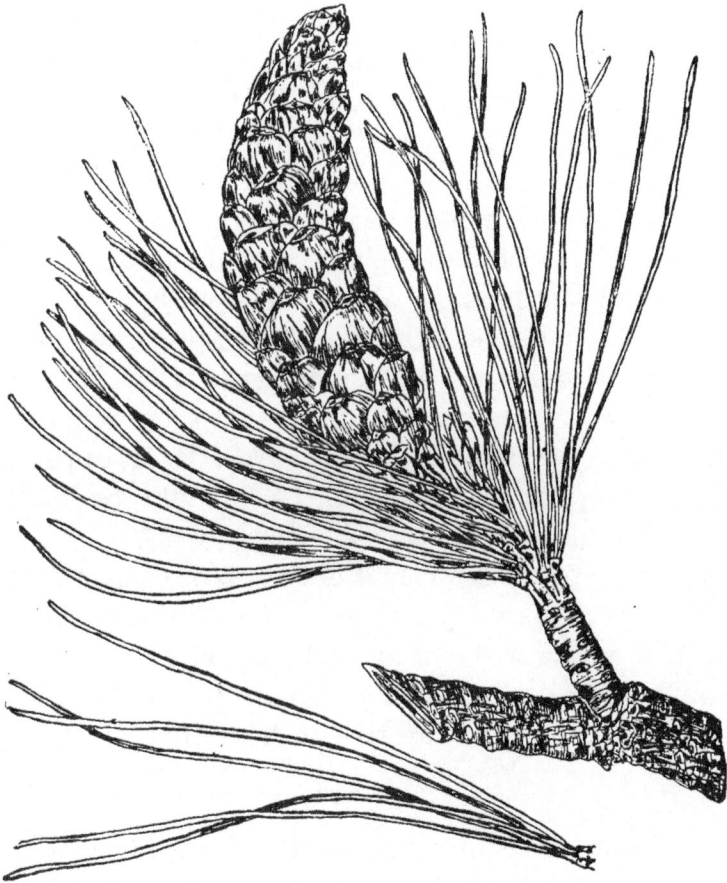

FIG. 51. SPRAY OF WHITE PINE WITH YOUNG CONE.
Below, a bundle of five leaves. Both nearly natural size.

The little catkins around the base of this young shoot carry the stamens. About the middle of June, when they are mature, they shed a yellow pollen and then wither and fall off, as we have observed. The small green bodies at-

Observe woodpeckers, chickadees, blue jays, hawks, and owls.

tached to the ends of young shoots are the staminate flowers, in which each young ovule is protected by a scale. For these we also looked in June and found them mostly near the top of the trees, where the boys picked them; but we found far more staminate flowers than pistillate ones.

From our observations made on other trees, we knew at once that the young cones, as the fruit of pine trees is called, had the pollen carried to them by the wind. Although, to judge from the abundance of pollen produced, the cones must have been fertilized, they grew much less during the summer than we had expected. But John thinks he has solved the problem about the slow growth of these cones. He found some twigs that bore these small cones at the tip, but bore mature cones with seeds at the base of the season's growth. He has concluded that these cones do not mature until the second autumn after their appearance. Examine carefully the material he brought and tell me what you think about his conclusion.

In the mature cones we find one or two seeds under each scale. These seeds are provided with a thin wing, by means of which they can sail considerable distances on the wind. Do you know at what time the mature cones open? Of what advantage is it to the tree to have the cones near the top?

§ 55. Besides the white pine, the Red Pine (*Pinus resinosa*) and the Labrador Pine (*Pinus Banksiana*) are common in the Northern States and in Canada, the latter often forming large forests. Farther south and along the eastern coast other species take their place. The range of the white pine is from Newfoundland to Manitoba, south along the Alleghenies to Georgia, and to Illinois and Iowa.

If evergreens are common in the neighborhood, the teacher ought to make himself and pupils acquainted with most of

them. Where conifers do not grow wild, very good material
is furnished by the Norway Spruce and the Scotch Pine,
which are often planted for ornament.

To the pine family belong also the Tamarack, the Spruces,
and the Firs. Spruces and firs are most commonly sold as
Christmas trees. The leaves of the spruces are set all
around the branchlets; those of the fir appear two-ranked
like the teeth of a comb.

If the Norway spruce or the Scotch pine are made the
basis of the lesson, material similar to that for this lesson
must be procured; and the teacher must work from a carefully constructed outline and must have directed the children to observe the trees while they are flowering. The
flowers of the Norway spruce appear in May; those of the
Scotch pine probably early in June. Where groves of some
kinds of pines are not accessible, the teacher may take a
Christmas tree for the object of the lesson, and give the
children some idea about pine forests by means of pictures,
which can be selected from different sources.

§ 56. Pine Forests and Forests of Deciduous Trees Compared.

MATERIAL: A walk to a grove of pines, if one is accessible. A collection of twigs, cones, and flowers of all the trees of the pine family
which the neighborhood affords; plants and dried berries of wintergreen; blueberry and other plants characteristic of the pine forest.

Some time ago we studied a dozen species of broad-leaved
trees; but on our excursions to the woods we also noticed
that the number of such trees in our neighborhood is considerably larger. And what a bewildering number of small
trees, shrubs, and woody climbers we found associated with
the larger trees! There grew the wild cherries, the wild
plums, the June berry, gooseberry, currant, raspberry, the
dogwoods, the hazel, the snowberry; also the wild grapevine, the false bittersweet, and several other climbers.

The pine forest seems to have taken us to a different world. In one region we find dense forests of almost nothing else but Labrador Pine. As far as our eye can penetrate we see the maze of gray, lichen-covered branches and tall, slim trunks. We are tempted to go on and on, hoping to come to the end of these silent and gloomy woods. All the trees are of nearly the same height, the same thickness, and show the same mode of branching. After we have walked for miles, we come upon a poplar thicket or upon a growth of younger pines; but they only mark the path of a conflagration which swept through the primeval forest. Now we see an opening in the thicket, and suddenly we stand on the shore of a sparkling lake. The tall, waving rushes, the broad-leaved lilies, the murmur of the rippling waters, the blue sky, with its white, floating clouds reflected from the glassy expanse of strength- and health-giving waters are the very impersonation of perfect rest, quiet, and happiness. And all around this sparkling gem grow the dark, melancholy pines. Their trunks are hoary with age, a bluish haze hangs over their tops; the eye tries in vain to reach the end of the forest. Lumbermen and hunters tell us that there is nothing but Jack Pine, or Scrub Pine as they often call them, for fifty miles around us.

Could we spend a month in the pine regions of our Northern States, we should also come upon large forests of grand White Pines or into airy groves of round-topped columns of Red Pines. Unless we were experienced woodsmen, we should, no doubt, lose ourselves in the dark Tamarack and Cedar swamps, where the trees stand only a few feet apart and where fallen trunks and deep holes burned into the ground by former forest fires make travelling almost impossible. These swamps, the terror of woodsmen and hunters, are the home and safe retreat of the moose, the grand monarch of our forests.

Wherever the pines stood very thick, little else but dead, brown needles and a few gray lichens covered the ground. In places less densely shaded the ground was covered with low bushes of blueberry, with a sprinkling of wintergreen, and a few other small shrubs; while the dense shade in Tamarack and Cedar swamps permitted no undergrowth except mosses and lichens.

We have, however, by no means exhausted the list of differences between pineries and broad-leaved forests. Were we to return to these solitudes in autumn, we should look in vain for the golden yellow of the ash or for the glowing tints of oaks, sumachs, and Virginia creeper; the pines still wear their sombre green of summer, but gray clouds, drizzling rains, and cold winds make the scenery still more melancholy and gloomy than it was in summer.

When, however, fierce snowstorms at last follow the autumn gusts, then the pine forest is suggestive of shelter, home, and life. The broad-leaved trees now look bare and lifeless, but the evergreen needles of the pines remind us that life is only sleeping, and conjure up the scenes of many a merry Christmas; and, while the blizzards may sweep over their tops, they are hardly felt on the ground.

The impression which the pine forest makes upon us by the uniformity of its trees, by the absence of bright flowers and luxuriant underbrush, and by its sombre foliage, is very much deepened by the almost solemn silence which prevails in it. The winged musicians, which love the leafy brush and its cool shade, find nothing to attract them here; bees and butterflies find but few flowers, and are, therefore, scarce; and only a few insects feed on the green pine needles. Rabbits, foxes, and wolves also find little to attract them in the deep forest, and are not as common there as most people think. The Virginia deer, however, is now almost restricted to the pine regions, at least in the

s

Northern States. It finds shelter in the thickets, but feeds mostly in the openings around lakes and grassy ponds. As all these wild animals have very acute senses of smell and hearing, they generally observe a man before he sees them and disappear into the thickets or lie unobserved by him.

§ 57. Importance of Pine Forests for North America.

Pines in the economy of nature.

MATERIAL: Show on a map the areas which are largely occupied by pine forests.

What we have learned about the powers of other forests to retain moisture is just as true of the pine forest, and under this term we include all trees belonging to the pine family. Pines are found especially in the northern part of New England, around the Great Lakes, about the head waters of the Mississippi in Minnesota and Wisconsin, in the wooded parts of the Rocky Mountains, and on the Pacific slope in Oregon, Washington, British Columbia, and Alaska. Who will tell me from the map what streams have their head waters in pine regions? An abundant precipitation falls in these regions and supplies the water for countless mills and factories; it fills the thousands of Minnesota, Wisconsin, and Canada lakes, and carries ships on the greatest system of navigable rivers and lakes in the world. The extensive deforestation, which has been going on ever since the United States began to be settled, is partly the cause of the disastrous floods which occur almost annually along the lower Ohio and Mississippi rivers. Were these regions entirely deforested, the spring floods would become still more disastrous, while in August and September boys would be able to wade rivers that had been a mile wide in April. Chittenden, in his book on the Yellowstone National Park, estimates on good authority that the forests in that region retard the melting of the snow at least six weeks, and a similar effect is exerted by forests everywhere, although it may not be so considerable. What rivers are fed by the streams from Yellowstone Park? Why are our pine forests of special importance as conservers of moisture?

Destruction of pineries. — Millions of acres of pine have been cut down by lumbermen, but they do not reseed the land cut over. They are interested only in the pine, which nature grew without any labor or expense on their part, and when that pine has been cut, they let the land revert to the respective states or counties for taxes, while their axemen in thousands move into virgin forests. However, the lumberman's axe would never destroy our pineries, for the forests would soon reseed themselves and they would flourish as before, if no graver danger were induced by it. *The great, the awful, destroyers of the pineries are the forest fires.* These fearful conflagrations, which almost annually sweep over hundreds of square miles, are caused in several ways.

The brush and dead trees which lumbermen leave soon become bone-dry under the summer sun, and furnish an enormous amount of the most combustible material. Fires which are built by lumbermen, hunters, campers, and Indians, and are carelessly left burning, may be fanned by the wind into widespread forest fires. The burning of brush by settlers, and especially the sparks from locomotives, are frequently causes of forest fires. It must not be forgotten that in the periods of drought which are so common a feature of our summers, a smouldering match thrown on the dead pine needles will almost certainly start a fire; and it is a sad fact, but nevertheless a fact, that the fools who do not think, and the brutes who do not care, are always with us. The awful extent of ruined forests which one sees along the railroads in the pine regions of Minnesota, Wisconsin, Michigan, and Canada shows only too painfully the fearful destruction human negligence has wrought.

Can 'forest fires be prevented ? — Many states have passed laws to prevent forest and prairie fires, and are making an earnest effort to enforce them ; but that it is difficult to punish offenders who commit crimes fifty or a hundred miles from the nearest settlement is easily appreciated. The writer is forced to believe that, unless all the states concerned speedily enter upon the work of a forest administration as scientific and as effective as that of the German States, Austria, and France, these terrible fires will continue to burn as long as there are forests left to feed them.

The teacher should impress upon the pupils that any man or boy who carelessly leaves a camp fire burning endangers human life and property, is not worthy to be a free citizen in a free state, but should be treated as a criminal or as insane.

See John Muir, The American Forests, Atlantic Monthly, August, 1897.

§ 58. Value and Use of Pine Lumber.

MATERIAL : Pieces of pine wood and hard wood ; rosin ; a bottle of turpentine ; a resinous piece of pine. If practicable, the pupils should also visit a sawmill and a lumber yard.

OUTLINE FOR LESSON

1. Nearly all the lumber used in the building of houses, barns, factories, sidewalks, etc., in the prairie states came from the pineries. The trees are cut in winter, hauled to the banks of streams on sleighs, and in spring they are floated down to the sawmill, and cut into lumber, which is distributed by the railroads over the whole country. The operation of logging, booming, and rafting should be described by the teacher, and illustrated by pictures procured from magazines or other sources.

2. Different kinds of pines are used for

Fuel,	Bridge piling,
Telegraph poles,	Railroad ties;

and in the manufacture of

Furniture,	Pencils
Railroad cars,	Matches.

The pines furnish us with more lumber than all other trees taken together; their wood is sufficiently durable for many purposes, and on account of its softness and straight grain it can be easily worked; their tall, branchless trunks are just the right material for long, straight timbers; for these reasons the pines are commercially our most valuable trees.

IX

A FEW BIRDS THAT ARE RESIDENTS IN OUR NORTHERN STATES. DECEMBER

All of you children are well acquainted with the House Sparrow, one of the few birds seen here all through the winter. There are, however, several birds which remain with us even in the most severe winters, although many of us may not see them frequently. Among these are several owls and hawks, the Blue Jay, a few woodpeckers, and the Chickadee. Inseparably connected with the forests are the woodpeckers. We shall first learn a little about the

§ 59. Hairy Woodpecker. *Dryobates villosus.*

MATERIAL: Stuffed bird or picture. Outdoor observations.

Every boy and girl has seen a bird now and then that climbed up and around trees by little jumps, and which at the same time seemed to examine the tree very carefully, and here and there tore off little pieces of bark. The most conspicuous of these birds is the Red-headed Woodpecker; but as the latter is not a winter resident with us, we shall study the characteristics of the family in the Hairy Woodpecker.

Description. — Both male and female are black and white above and have a white stripe on the middle of the back; the male alone has two red spots on the back of the head. The under parts are grayish-white. The bird is from nine to ten inches long, being nearly the size of a robin.

Observations. — If you can get out into the woods, look for the tracks of minks, foxes, weasels, and raccoons.

Let us see now how the hairy woodpecker is adapted to the life it leads.   A woodpecker's foot differs from that of a sparrow by having two toes pointed forward and two backward.   The toes are provided with long, curved claws, by means of which they can hold very firmly to the bark of trees.   Our woodpecker generally begins to examine a tree

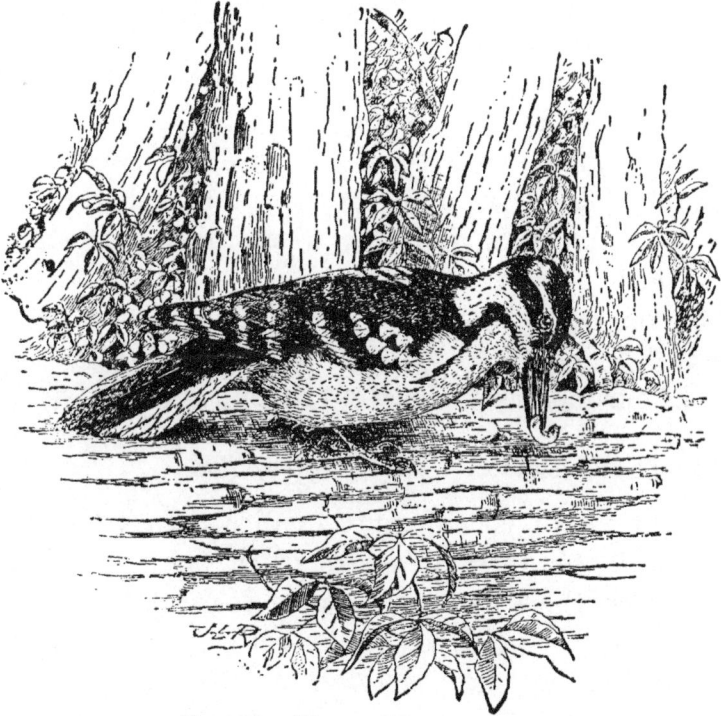

FIG. 52.   HAIRY WOODPECKER.

quite low on the trunk and then travels upwards.   Were he to climb downwards also, he would have to depend entirely on his feet, but when going up he uses the stout feathers of his short tail as a kind of rest or brace.

His food. — He does not examine the trees for fun, but is hunting insects.   His strong beak enables him to break

Visit a collection of mounted North American mammals, if such collection is within your reach.

off pieces of bark, and even to cut holes into partially decayed wood; *and in this way he finds many insects which are out of reach of most birds.* His tongue is awl-shaped, and provided with a barbed, horny tip, which enables him to hook the worms in their holes.

Other common woodpeckers are the small Downy Woodpecker, the Red-headed Woodpecker, and the Flicker; the last two are migratory.

All woodpeckers nest in holes in trees, which they hollow out themselves. Sometimes they catch insects on the ground, but their movements there are awkward. All woodpeckers are useful birds; their nests are always built in dead trees, or in the dead wood of living trees; no boy should ever molest them.

Could you tell a woodpecker from a pigeon by its mode of flight?

One of our most common winter residents is

§ 60. The Chickadee. *Parus atricapillus.*

This little bird, considerably smaller than a house sparrow, is often seen in small flocks, or in pairs, busily climbing about on the twigs and branches, carefully searching for concealed insects, pupæ, and insect eggs. They are best identified by their call note, chickadee, day, day, day. Above, they are gray; below, almost white; the crown, nape, chin, and throat are black; the bill and feet lead-black. The nest, which is placed in the holes of stumps, consists of wool, fur, feathers, and other soft material. The eggs, which may be looked for in May, are white, thickly sprinkled with brown. The song is very simple, sounding like chickadee-dee-dee-dee. Although they are with us the year around, you are likely to notice them most in fall and spring, when they are most numerous in groves and gardens.

§ **61. The Blue Jay.** *Cyanocitta cristata.*

Material : Picture, mounted bird, old nest.   Outdoor observations.

This is our most noisy and most showy winter resident. He is indeed a beautiful bird; but it must be said that his heart is not as good as his dress is splendid. You cannot fail to identify the blue jay. He is a little larger than a robin. The color of both male and female is blue above; the head is adorned with a fine crest; the wings and tail are a bright blue, barred with black. The song is a clear

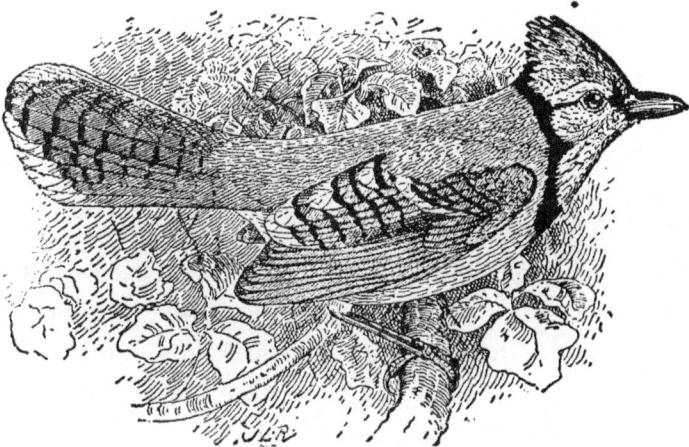

Fig. 53.   Blue Jay.

whistling, bell note, sounding like deedle-léet, deedle-léet ; but it is quite loud, and does not at all resemble the feeble chirps of the warblers. Its loud scream, jay, jay, jay, is heard much more commonly, and sounds very cheerful on a crisp, winter day. The nest is built in small trees from ten to twenty feet above the ground; it is made of small sticks, coarse roots, and twigs, and lined with finer material and leaves. The jays seem to be most common in oak woods, and often build quite near to houses.

Food. — The jays eat anything which a bird can eat: seeds, buds, acorns, frogs, etc.   During the breeding season, they destroy some young birds and eggs; but recent investigations by the Department of Agriculture show that this habit is not nearly as common as it was believed to be. Of two hundred and ninety-two stomachs of jays examined, only two contained remains of young birds, and three contained shells of small birds' eggs.

While it might not be wise to allow many jays to nest near our homes, they evidently do not deserve to be generally persecuted. They eat a great many injurious insects, do practically no harm to agriculture, and they enliven our winter landscapes more than any other bird.

Hawks and Owls

We have learned that most mammals live on vegetable food; a few, like bats, shrews, and moles, live largely on insects; while others live mostly on the flesh of mammals and birds. We shall now study a few birds which do not feed on seeds and insects, as most birds do, but which live largely on birds and mammals, which they catch with their sharp talons. I think we shall be able to understand readily that hawks and owls, the birds I refer to, are well adapted to catch and devour birds and rodents, but that they could not possibly pick up wheat and seeds as chickens do.

§ 62. The Red-tailed Hawk. *Buteo borealis.*

MATERIAL: A picture, or a mounted bird; this, or some other hawk, caged; a freshly killed specimen. Outdoor observations.

This is one of our large hawks, which every child has seen sailing high in the air. There they soar so easily, seldom moving their wings, that we cannot help thinking we could do it too, if we only had some kind of wings. But up to this time, it must be confessed, all wings and flying machines invented by man have proved very dangerous contrivances.

Description. — The Red-tailed Hawk is not really a winter resident in Minnesota, but it returns from the South about the first of March.   The nests are built early in April, in large

FIG. 54.   RED-TAILED HAWK.

trees in the woods. The material used is coarse sticks, twigs, leaves, and mosses. The nest, on account of its size, cannot be concealed, but its elevation renders it a fairly safe place. The color of this bird is a dark brown above, with many bars and streaks; *the tail of the adult is rust-red, with a narrow black bar near its end.* The male is about twenty-two, the female about twenty-four, inches long.

Food and habits. — When you see these birds soaring leisurely over a wide, deep river bottom, they are either out for an airing, or their keen eyes are spying for young rabbits, wood rats, mice, and other small rodents. When they espy their prey, they pounce down upon it, fasten their long, curved talons in it, and carry it off to some convenient place, where they tear it to pieces with their hooked bill. They do not always distinguish poultry from wild birds; and most farmers shoot them wherever they can, although it has been shown that eighty-five per cent of their food consists of small, injurious rodents. They should, therefore, not be molested, unless they habitually visit your barnyard.

You may now compare the structure of the red-tailed hawk with that of the duck, and show how each bird is adapted to the life it leads. If illustrative material of some other hawk is more accessible, the teacher may substitute that. Any one of the larger or of the smaller hawks shows the structure that is typical of birds of prey.   •

§ 63. The Great Horned Owl. *Bubo Virginianus.*

MATERIAL: Similar to that for hawks.

A small boy was out camping in the woods for the first time in his life. Towards evening a terrible thunderstorm swept through the woods. The thunder crashed incessantly, the storm roared, the trees were bent, the little boy looked scared and asked his father if the tent was going to fall down on them. Half an hour later it was dark, nothing could be seen, but many things the small boy heard, which made him feel ill at ease. Suddenly some one called out right above the tent, Hoo-hoo-hoo! Waugh-hoo! That was too much. Some goblin, ghost, or Indian was surely

FIG. 55. GREAT HORNED OWL.

looking for him; his tears began to flow and, with the most plaintive voice, he cried, " I want mamma!"

There were no Indians within a hundred miles of the tent, but the Great Horned Owl was sitting in a tree above it, and his hooting had broken down the little fellow, who had talked fishing, hunting, and camping for the last two weeks and had bragged not a little to his less fortunate play-mates.

Description. — This owl reaches a length of from twenty to twenty-four inches, and its loose, fluffy feathers make it look bigger than it really is. Out of this mass of feathers project a vicious-looking crooked bill, and formidable curved claws. Two large tufts of feathers stand erect on the cat-like head, and the large yellow eyes seem to be intent upon staring you out of countenance. The plumage is irregularly mottled and barred, showing shades of black, tawny, buff, and whitish; the feet and legs are also feathered.

This owl is a resident of our Northern States; it ranges from the Atlantic coast to the Mississippi valley and south to Central America. The nest is usually built on high trees in a dense forest. It is a rude structure built of sticks and twigs, and lined with leaves, moss, and grasses. The eggs, from three to four in number, are yellowish-white.

That this owl is a bird of prey, we perceive at once from the shape of its bill and claws. There are but few birds which it will not attack, and such small mammals as rabbits, rats, and mice are its daily fare. Should it chance upon the chicken trees of some careless farmer, it is sure to cause sad havoc, perhaps killing half a dozen chickens in one night. Where poultry is protected, it is undoubtedly a useful bird, like nearly all our birds of prey, because it helps to keep in check the army of small rodents which are nearly all more or less injurious to man.

Did you ever stop to think why the eyes of owls are so large? If their eyes were like ours, they could not find any mice after sunset. Their hearing also is very acute, and their flight, on account of their soft feathers, is noiseless. *The three points just mentioned are of great importance to owls, because the mammals on which they feed are mostly noc-*

*turnal, move about almost without noise, and having very sharp
ears they would hide at the slightest flapping of wings.*

It is very much to be regretted that the custom of indis-
criminately shooting all hawks and owls, and nailing their
dead bodies to barn doors, does still prevail in this country.
All the smaller hawks and owls are beneficial birds, feeding
almost entirely on insects and mice; some of the larger
ones are harmless; only a few, such as the Goshawk,
Cooper's Hawk, and the Sharp-shinned Hawk, may be truly
classed among the harmful birds.

REFERENCES ON BIRDS

1. Farmers' Bulletin, No. 54, Some Common Birds.
2. Yearbook, Department of Agriculture, 1894, Hawks and Owls as
   Related to the Farmer. Every teacher should secure both articles
   and refer the older pupils, especially the boys, to them.
3. *C. C. Abbott.* Birdland Echoes. J. B. Lippincott Co. A very
   interesting book.

§ 64. Review of the Birds.

MATERIAL: A boiled egg; some large tail and wing feathers; the
down and breast feathers of a duck or goose; pictures or mounted
birds representing the groups mentioned.

The covering of all birds consists of feathers, which grow
from the skin and are in no way connected with the bones.
Birds do not keep the same covering of feathers during their
whole life. Once or twice a year the old feathers are shed
and new ones grow rapidly. This process is called moulting.
During this time birds do not sing, and their usual activity
is much reduced. The males of most birds possess a much
brighter plumage than the females, and in general the plumage of birds is brighter in spring than it is in fall and winter.
Let the children mention a few birds to which this statement is specially applicable.

You must have been struck by the great resemblance in
shape and general appearance that one bird bears to the

other. This resemblance is due to the fact that all birds have their fore limbs changed into wings, are covered with feathers, and have a horny bill. Do you know of any birds that cannot fly?

All birds lay eggs, and nearly all in nests more or less skilfully built by themselves; only a few lay their eggs in the nests of other birds. The eggs consist of a hard, calcareous shell, the white, and the yolk, which is of a yellow or reddish color. Every egg has a small space filled with air.

The warmth of the brooding bird causes the young to develop in the eggs. If during the period of incubation the eggs are thoroughly chilled, the young birds in them will die. The time necessary to hatch the eggs varies much with the size of the birds. The young of most birds are blind, naked, and helpless when hatched, and depend entirely on their parents for food; but the young of chickens, ducks, and geese are covered with a soft down, can see at once, and after a day or two they move about with the mother and find their own food.

Most birds breeding in cold or in temperate regions migrate south on the approach of winter and return in spring; these are *Migratory Birds*. A bird that remains in the same region winter and summer is called a *Resident Bird*.

The most characteristic marks for the classification of birds are found in the structure of the bill, the legs, and feet.

A very large group of birds whose feet are adapted to perching on trees, and most of which have some power of song, are frequently called *Song Birds*.

A much better marked group is formed by ducks, geese, swans, and similar birds. Their body is somewhat boat-shaped; their legs are short and are set well back; the toes are webbed and form powerful paddles. They can use their bill as a strainer to separate small aquatic animals from the water.   Their plumage is always kept well oiled, so that it is not wetted, although they swim and dive for hours. These birds are called *Swimmers*.

Another well-marked order is formed by the *Waders,* or *Marsh Birds*.   Their legs are very long and enable them to wade about in shallow water.   Their neck and bill are also long and enable them to procure fish and other aquatic animals.

You may now try to find some of the characteristics of *Woodpeckers, Chickens,* and *Birds of Prey*.   How do owls differ from hawks ?

X

ABOUT A FEW COMMON MAMMALS.
JANUARY

§ 65. The Common Skunk.

MATERIAL : Mounted specimen ; skin, picture, skull.

The range of the common skunk extends from Hudson's Bay to Mexico. Its color is quite variable, but is usually black with a white patch on the nape, white stripes on the back, and a white tip on the tail. The animal generally moves about slowly, bears its bushy tail erect, and does not try hard to run away from man unless very near to a place of concealment. Compared with the slim, almost snake-like body of minks and weasels, it reminds one of a small, fat dog. Its whole length is about twenty-eight inches; the tail measures nine inches. It is best recognized by its horrible, mephitic odor.

Habits and food. — The feet of skunks are provided with strong claws for digging. Their burrow extends in a straight line for about seven feet and about two feet below the surface. Then it ends in a large excavation. This is the hiding-place of a skunk family, and here they remain inactive during the winter. They do not lay in a store of provisions, but are very fat in the fall, and to fast from December to February or March seems an easy task for them. Like the

---

Observations. — Visit a zoölogical garden, if you live in a large city, and learn to know the Virginia deer, the elk, the moose, the black bear, the grizzly bear, the wolf, the coyote, the wild cat, the lynx, the puma, and the buffalo, or bison.

raccoon and other animals, they do not hibernate in the Southern States.

The food of skunks consists of young birds, eggs, frogs, mice, insects, and any small creatures they can catch on the ground. If they live near a farm, they frequently draw upon the farmer's poultry for a supply of meat and eggs; and being rather slow and stupid animals, they are often caught by dogs, shot by the farmer's boy, or even clubbed to death. The only effective means of defence they have is their terrible squirt-gun. The foul fluid is secreted and emitted by two special glands situated near the root of the tail; it is not the urine as some people think. The skunks never soil themselves with this fluid; they are very cleanly, and often no smell is perceptible near the mouth of a burrow in which a whole family is living. Their black and white colors and their bushy tail are probably a warning to the larger beasts of prey, and thus protect the animal by making it conspicuous.

§ 66. The Common Mink. *Putorius vison.*

MATERIAL: A mounted mink; skin or picture.

Description. — The color of this well-known farmer's pest is a uniform chestnut-brown, not appreciably lighter below; but the tail is darker, the end of the chin is white, and there is also a white spot between the fore legs. The general appearance of the mink is that of a long, slim cat, with a somewhat bushy tail which is from six to eight inches long. The head and snout are more elongated than those of a cat, and the animal shows its amphibious nature by having all the toes webbed; it measures about twenty-eight inches from the nose to the end of the tail bone.

Habits and food. — The mink, like other small mammals, has found it advantageous to live near the abodes of man.

All these animals can be seen in Lincoln Park, Chicago.

A farmer who lives near a small stream and a few acres of timber or brush is likely to have a large number of chickens killed by this marauder, which will make its home in the crevices of rocks or under stone and wood piles, from where it makes regular excursions to its hunting-ground, the farm-yard and outbuildings. Minks are, however, not as bad as weasels, for the latter often kill ten times as many chickens as they can eat. In the woods minks live on eggs, birds, rabbits, mice, and other small animals; and as they are good swimmers, they also catch fish.

The mink is one of our fur-bearing animals; it is found over the whole of North America, and thousands of skins are annually used in this country and in Europe.

It belongs to the Weasel family; its near relatives, the Pine Marten, the Weasels, and the European Ermine, all have long, slim bodies and emit an unpleasant odor.

§ 67. The Red Fox.

MATERIAL: Mounted specimen ; skin, picture, skull.

This fox inhabits the whole of North America as far south as Texas. The Cross Fox and the Black Fox are considered as varieties of the Red Fox.

Every boy who has ever seen a fox will tell you that he looks like a rather small dog with a bushy tail. The general color of this fox is a reddish-gray, the feet and ears are black, the tail is tipped with white, and there is a narrow line of dull white on the belly. The whole length of the animal is about forty-five inches; the tail alone measures about fifteen inches.

Habits and food. — This species, like all foxes, digs holes in which its young are reared and to which they occasionally retreat when hard pressed by pursuing men or dogs.

Observations. — If you cannot see the animals themselves, learn to distinguish them on pictures, or examine mounted specimens.

It is probable that even foxes are at least as numerous now as they were before the country was settled.

They live on wild and tame birds, rabbits, muskrats, squirrels, mice, and other small animals. When they hunt birds, they either lie in wait for them or they trail them like a pointer dog and then suddenly spring upon them.

The different kinds of foxes are among the most valuable of fur-bearing animals. Their cunning is proverbial, and it is difficult to shoot or to trap them. Newhouse, in his "Trappers' Guide," recommends to smear the trap and chain with blood or to give both a coating of beeswax.

§ 68. The Raccoon.

MATERIAL : Like that in preceding lessons.

OUTLINE FOR LESSON

Description. — About as large as a medium-sized dog, but legs shorter; weight about twenty pounds; body grayish above mixed with black; ears and lower parts whitish; *a black spot on the cheeks; tail with alternating black and gray rings from ten to twelve in number.*

Habits and food. — Their nests and winter retreats are in hollow trees; they are good climbers. They feed on clams, birds, eggs, turtles, frogs, and are very fond of corn in the milk. Their flesh is eatable and they make very interesting pets. In the Northern States they hibernate for several months.

XI

AMERICAN BIG GAME. FEBRUARY

We have observed quite a number of birds and mammals and among them a few that are hunted or trapped for sport or profit. Who will name a few of these animals? We have frequently had occasion to note that most birds and smaller mammals prefer a settled and partially cleared country to the wild primeval forests. If the smaller mammals were taken only in the season when their fur or their meat is good, we should always have plenty of them around us; but they must be kept in check, for in large numbers many of them would become a great pest to the agriculturist. When, however, we find them in places where they do no appreciable harm, no boy will kill them for the mere love of destruction.

Now I will tell you about a few large animals whose home has always been in the large forests and who have disappeared from all the thickly settled regions of our country. Probably you have all heard about the graceful Virginia Deer, the stately Elk, the majestic Moose, and also about the Wolf and the Black Bear.

§ 69. The Virginia Deer, or Red Deer. *Cervus Virginianus.*

MATERIAL : Picture ; mounted heads ; antlers ; visit to museum or zoölogical garden.

This beautiful animal was formerly very common from Maine and southern Canada to the Rocky Mountains, and

---

Note. — Give the pupils some good reading along the line of Nature Study. See Appendix.

south into Texas and Mexico. It was not only found in the large forests, but was also found in the seams of timber which border the streams and surround the lakes in the prairie states. It is still not rare in the sparsely settled regions of the Alleghanies, the New England States, and in the woods of northern Michigan, Wisconsin, and Minnesota. Of all our large game it holds out longest against the advance of civilization, and if it is properly protected and city sportsmen learn to control their hunting passion and shoot only one or two in a season instead of killing them by the wagonload, there will be plenty of deer for a long time to come ; if not, then the Red Deer will soon follow the Moose and the Elk into almost inaccessible regions, and finally by the brutality and stupidity of *enlightened American citizens*, the sad fate of the Bison will overtake it. But if the boys and young men now in our schools and colleges will develop as much common sense as the writer sincerely hopes and believes they will do, then this can never happen. *Monarchial Europe has preserved much of her big game by severe and often cruel laws; republican America ought to be able to do more by the intelligence of her citizens.*

Description. — A full-grown Virginia buck is about as large as a yearling calf, but you must not forget that its head is thinner and its muzzle much more pointed; that its legs are longer and only about as thick as those of a large sheep. The fawns are at first bright reddish-brown and spotted with longitudinal rows of white, but after a few months the spots disappear. Then the general color of bucks and does is chestnut-red in summer, changing to grayish in winter. The chin, throat, abdomen, and under side of the tail are white. Their color blends so well with the brown of dead ferns, pine needles, and other foliage that it is almost impossible to discover them at the distance of a hundred yards, unless they move or are in the open. Their

ability to hide in brush and grass is also remarkable.   The
writer and a friend once nearly failed to discover a flock of
eight or nine tame deer that were lying in a field of very
thin timothy which was only about a foot high and where
there was no brush of any kind.

FIG. 56.   VIRGINIA DEER.

The bucks alone are provided with antlers.   The first
pair are mere spikes, but they increase in size and in the
number of branches until the buck is about five years old,
when each antler has about four prongs.   In January or
February, according to latitude and season, the buck sheds
his antlers, and after three or four weeks the new antlers
begin to grow.   They are covered with skin and a velvety

growth of hair; are very soft and will bleed freely if scratched or otherwise injured. At this time the buck is not combative; when he must fight he uses his fore feet only, and it is almost amusing to see how carefully he brushes away blood-sucking flies and mosquitoes from his tender antlers.

Habits and food. — The food of this deer varies much with the season. In spring and summer they feed on tender grasses and herbs around lakes and ponds, and also eat cultivated plants in fields and gardens. In the winter time they subsist on the buds and twigs of a great many shrubs and trees. On warm days they feed mostly during the cool hours of morning and evening, while during the remainder of the day they lie concealed in thickets or in the tall grass of swamps. When surprised by man, they rush into a thicket, from which the old bucks occasionally watch their enemy for ten or fifteen minutes, stamping with their feet, and uttering a sound which is half-way between a snort and a whistle. The antlers of the bucks have attained their full growth and have become hard in August or September; the skin on them has dried up, and is brushed off by rubbing the antlers against trees and bushes. In November the rutting season begins, and the bucks move about constantly in search of the does, and fight fierce battles with one another. Sometimes the horns of two stags become locked, and they starve to death, or fall a prey to wolves. Audubon relates a case where the skulls of three deer were found with their antlers firmly locked.

The natural enemies of deer are wolves, the Canada lynx, the wildcat, and bears. The latter probably seldom catch a healthy deer; the lynx lies in wait for them on the ground or on trees, and springs upon them in cat-fashion; and the wildcat kills, no doubt, many a fawn. Wolves frequently hunt a deer in packs. In the summer time the hunted deer takes to the water, if possible, and will readily swim any large river or a lake a mile wide and escape; but hunters say that, in the winter time, wolves nearly always run down the deer which they start to chase.

Economical value. — The meat of the red deer is the most palatable of all kinds of venison. The pioneer settler, who lives on venison and little else for years, the lumberman, the hunter, the Indian, and the guests of our luxurious metropolitan hotels, all prize it. Its skin furnishes valuable robes and leather for Indians and whites, its antlers furnish the handles for much fine cutlery, and the fine mounted heads adorn the home of many a sportsman. Deer do no injury unless they make frequent inroads into a settler's field and garden.

Hunting the deer. — The true sportsman gives his game a fair chance to escape. His endurance, skill, and knowledge of woodcraft are pitted against the fleetness and endurance of his game. He finds his deer without the assistance of dogs, which he can do best by tracking them after a light fall of snow. He prefers bucks to does, *never shoots at a fawn, never kills more game than he can make use of, and obeys the game laws of the state in which he hunts.* Hunting with hounds is generally prohibited because they drive the deer into lakes and rivers, where they are simply butchered by men on land or in boats, and where they have been killed with clubs. The so-called fire-hunting is not sportsmanlike, because it does not give the deer a fair chance. It is likely to cause more game to be killed than the hunter can use. He cannot see whether he shoots at a buck, a doe, or a fawn; and in settled regions may shoot domestic animals or even people. The light of his lantern in the boat is reflected only from the eyes of the creature, which curiously gazes at the uncommon sight from the near shore. A true hunter is also merciful. He does not shoot unless he has a fair chance to kill, and he tries his best to capture all the animals he wounds; for a wounded deer is almost sure to wander about for days in agony until the wolves end its misery.

§ **70. The Elk, or Wapiti, and the Moose.**   *Cervus Canadensis* and *Alce alces*, var. *Americanus.*

FIG. 57.   ELK, OR WAPITI.

MATERIAL : As under Virginia deer.   It would be beyond the plan and purpose of this book to give a detailed description of the life and habits of the elk, moose, wolf, and black bear.   For such treatises teacher and pupils are referred to the many books on hunting big game ; to Audubon and Bachman, Quadrupeds of North America, and to various sporting magazines, especially to Forest and Stream, of

New York. Read the articles on the Yellowstone Park in Forest and
Stream, 1894.

*The object of this chapter on " Big Game " is to interest the
pupils in the grand and beautiful, as expressed in our large,
wild mammals, and to cause them to take an active interest in
their intelligent preservation, wherever conditions permit it.*
Compare the chapter on "Domesticated Animals."

Both the wapiti and the moose are deer. The wapiti,
often called elk, stands about as high as a horse. The
head of the male, called the bull by hunters, is adorned with
a pair of formidable antlers, each generally having six
prongs. The prongs that diverge most have points about
three and one-half feet apart, and a pair of the horns weigh
from thirty to forty pounds. The color of this grand and
beautiful deer is a chestnut-red in summer and grayish in
winter. Formerly the wapiti was found from Virginia to
the Rockies, but it is now very rare on this side of the
Rocky Mountains. It is estimated that about thirty thousand of these noble beasts roam in the Yellowstone Park,
where no rifleman may molest them, but where a lover of
nature may hunt them with a kodak.

The moose, which is the largest of all our deer, fully
reaches the height of a horse, but the maximum weight of a
very large bull is not more than fifteen hundred pounds.
The bull carries a pair of flattened antlers of enormous size,
weighing from fifty to seventy pounds. The head of the
moose resembles that of a horse, but the upper lip is considerably longer than the lower and enables the moose to
browse on twigs and peel off the bark of trees. Its ears are
about as large as those of a domestic cow or mule. More
than one inexperienced hunter has not fired at a moose cow,
because he was under the impression that some farmer's
mule had strayed rather far into the woods, and later the
same inexperienced hunter has killed the first mule he came

across. Unless we think of the moose in connection with
its wild and weird woods, we can hardly consider it beautiful. Its color is tawny above and somewhat yellowish below. It is still found in the wild woods of northern Maine
and Minnesota, in Canada, the Rocky Mountains, and on the
Pacific coast from Oregon to Alaska.

FIG. 58. MOOSE.

§ 71. Wolf and Bear.

MATERIAL: Similar to that for deer; children might see some of
these large animals in the circus.

Wherever red deer, wapiti, and moose are found, you
may also come upon the wolf (*Canis lupus*) and upon the
black bear (*Ursus americanus*). Children frequently imagine
the beasts of prey to be very much larger than they really

are. Our common wolf is as big as a large dog. Its color is chiefly gray, but in Florida wolves are black, in Texas red, on the prairies, dusky. Wolves eat any animal they can catch and kill; they rarely attack man unless they are driven to it by hunger. They are exceedingly wary and are still not rare in many well-settled regions.

The black bear stands about three feet high, attains a length of about six feet, and weighs from two hundred to four hundred pounds. Bears live on the animals they can catch, but are also fond of acorns, wild cherries, and all kinds of berries; they also eat insects and rob the nests of wild bees' honey. A black bear very seldom attacks a man, unless a hunter wounds him, or a she-bear believes her cubs to be in danger.

§ 72. Review of the Mammals.

The young of mammals are born alive; they are not hatched from eggs. Their first food is the milk of their mother. This way of feeding young animals is a decided improvement upon the way in which birds feed their young. The mother can generally procure her food without much difficulty; and her milk furnishes food and drink at all times available for the young animal. The young of small mammals are always well concealed; those of the larger herbivores, like deer, cattle, and sheep, can run about a few hours after birth, and are protected and defended by their mothers; while the young of the large flesh-eaters have, probably, no enemies, except man. The eggs of birds are exposed to many dangers for a long period; but these dangers are avoided in the world of mammals. All animals whose young are born alive and are fed on the milk of the mother are mammals. Seals, dolphins, and whales are mammals, although the last two look very much like fishes.

Most mammals have a covering of hair, which is shed

regularly, and renewed in the fall and spring. The summer coats are much thinner than the winter fur, and only the latter has any value as peltry.

The different families of mammals bear no such strong resemblance to each other as we found among the different families of birds. Very conspicuous differences are exhibited in the shape of the limbs and in the dentition.

The feet of horses, cattle, sheep, hogs, and deer end in hoofs. These are therefore called *Ungulates*, or *Hoofed Animals*. Among them we easily distinguish the omnivorous hogs from the strictly herbivorous deer, cattle, sheep, and the one-hoofed horses. Both deer and cattle are ruminants, but the latter are easily distinguished from the former by their permanent horns.

Of those mammals whose feet are provided with claws, dogs, wolves, foxes, and the different members of the cat family have long, pointed eyeteeth, or canines, and sharp, pointed molars. These molars are well fitted to shear meat and crush bones, but they could not grind up hard seeds and herbs. Animals with such dentition live on the flesh of warm-blooded animals, *e.g.* on birds and mammals. They are not cruel or bloodthirsty in the real sense of the word; for they have to kill their prey, or starve themselves. These animals we call *Flesh-eaters*, or *Carnivores;* and the large ones are often called *Beasts of Prey*. The flesh-eaters prevent small and large rodents and all herbivores from unduly multiplying, and compel them to exercise their limbs and their wits. Whenever animals have been found that had no enemies, they were always very dull brutes, and were soon exterminated by man. The carnivores are therefore by no means superfluous or injurious in the economy of wild nature, although they may be injurious, or even dangerous, to man. A small order of mammals, to which the Bats belong, are *Insectivorous*.

Now you may try to find for yourself the distinguishing marks of the order of Rodents, or Gnawers. Of this important order Coues writes as follows: "Though a feeble folk, comparatively insignificant in size and strength, they hold their own in legions against a host of natural enemies, rapacious beasts and birds, by their fecundity, their wariness and cunning, their timidity and agility, their secretiveness, each after the means by which it is provided for exercising its instinct of self-preservation, among which insignificance itself is no small factor."

XII

SUMMARY OF LIFE IN THE WOODS

§ **73.** In no other place do we find so many plants associated as we do in the woods. There are, each in its favorite soil and locality, the many large trees; under them and near them grow the host of shrubs, still more numerous than the trees. On trees and shrubs nature's own drapery of wild grapevine, Virginia creeper, and other climbers and twiners is gorgeously displayed. Most shrubs and woody vines, no doubt, seek the company of large trees on account of the good soil and protection from withering winds to be found there; but such delicate flowers as White Heart, or Dutchman's Breeches, Bloodroot, Trillium, and many others, could not endure the glaring light and the heat of the sun, nor the hot, dry winds; the shade of the protecting foliage of trees is absolutely necessary to their existence. If you transfer any of these shade-loving plants to a sunny spot in your garden, they will soon die. The teacher might ask pupils to bring specimens of all plants to be found in the nearest grove. It is not necessary that pupils or teachers should know the names of all of them.

We have learned that the woods are a great plant community, in which the larger and stronger members protect and shelter the weaker ones.

This great plant community attracts a rich animal life although we saw little of this except song birds and insects.

Observations. — Look in open streams, brooks, springs, etc., for aquatic insects, fishes, hibernating frogs, crabs, snails, and aquatic plants.

Many of the mammals are nocturnal, and all of them find places of concealment, which shows that the woods are the favorite abode of many animals.

The demands animals make upon the woods vary as much as the animals themselves. Bees are attracted by the flowers of plums, cherries, and linden; butterflies are in quest of leaves for their caterpillars; beetles and other insects go there to find their favorite food; the birds are there in search of the insects and wild berries. The frog squats low to catch the fly, and the hungry snake glides about to find a frog. Large and small game find an abundance of food and shelter against the weather and against their enemies. Can you tell now why nearly every kind of carnivorous animal also dwells in the forest?

Thus it is seen that the woods are the storehouse of food for the animals living in it.

Willows, dogwood, grapevine, linden, and other plants attract insects to their honey, and the insects cross-fertilize their flowers. Birds and squirrels render important service to many plants by distributing their seeds, and by destroying injurious insects.

Many animals are important factors in the plant life of the forests, but they are not all beneficial to it.

In June, 1897, caterpillars defoliated the forests in the eastern part of Minnesota to such an extent that most of the trees looked as they do in January and February,—not a trace of foliage was left on them.

Life in the forest shows a marked decrease during the winter months. All the plants are resting; some mammals, the frogs, toads, lizards, and snakes, and all the insects, are hibernating; and nearly all of the birds have left for more genial regions. But the warm sunshine of spring and the April showers awaken nature to new life.

During the summer months there are more signs of life

during the day than there are at night. Bats, rabbits, mice, a few birds, and a number of insects are most active during the night; but most birds, bees, grasshoppers, the greater part of our butterflies, and most squirrels, are distinctly diurnal.

§ 74. Forests in the Economy of Nature.

We have alluded a number of times to the important part that forests play in the household of nature; we will now sum up and recall what we have learned about that subject. Let us try to recall what we saw when a heavy shower caught us last summer in the woods, and when we had to stand under the trees until the rain was over.

You remember that the roads and the cornfield were so dry before the shower that the dust was blown about on them. In the woods, however, no dust was flying, and the air felt very much cooler than that which blew from the south across the cornfield which we had passed; nor did we find any hard and baked soil such as we had observed all along the road. In well-shaded places we discovered tufts of maidenhair, and Bertha and Sadie dug up several tufts to plant under their porches. You remember that it had not rained for three weeks, and still the soil in which the ferns grew was quite moist.

The many fine roots, the humus, and the dense shade had conserved the moisture in the woods much longer than it could be retained by the exposed roads, fields, and prairies.

Before we had time to return home the sky became clouded, then the leaves began to rustle; a little later we heard a roaring sound far off, but it seemed to draw nearer; a few minutes later the storm rushed through the tree tops overhead, ashy clouds swept towards the northeast, and it seemed to be getting dark. We had found shelter under low trees with very dense branches and foliage. Suddenly a flash of lightning appeared to be shot perpendicularly from the clouds; a few seconds later a crash of thunder echoed and reëchoed through the forest. Then it began to rain. At first only single drops came through the leafy roofs above us, and it had rained about fifteen minutes before we were getting wet. On our way home, however, we found everything not only wet but thoroughly soaked, and leaves and grasses were loaded with large drops. Do you remember how, an hour later, the boys amused themselves with shaking the rain from the smaller trees? When we passed through the field, however, the corn, the wheat, and the grass were almost dry, and the roads were no longer muddy. The next forenoon some of the boys went into the woods again after flowers and found the brush still very wet; but on roads, fields, and prairie the plants and the surface of the soil were perfectly dry, although the rills made by the running water were still plainly visible. Did we see such rills in the woods?

From these observations we deduce the following: *In the forests the rain reaches the ground and soaks into it quite slowly; it does not fall upon bare soil; it forms no rills; a considerable part is retained on the foliage of trees and brush, and this part evaporates slowly and thus cools the air.*

Whether the water evaporated from forests has any appreciable effect upon the rainfall of the region is still a question under dispute, but is answered in the affirmative by some good observers. It is, however, an undisputed fact that treeless regions are more frequently visited by severe storms and cyclones and by hot, scorching winds. The great sources of moisture for our Eastern States and for the Mississippi basin are the Atlantic, the Gulf of Mexico, and the Great Lakes. As far as local showers are concerned, it seems to be a fact that they generally follow the wooded valleys and banks of rivers, or hover over lakes; but the writer is not able to give a satisfactory explanation of the phenomenon.

What becomes of the water which soaks into the ground? At the foot of a hill we found one large and a great number of small springs. All the springs united to form a little stream. The boys told us that the springs and the stream never dry up, and that their size is but little affected by those droughts which dry up many small prairie streams. We walked down the little stream to the large creek, and in both we observed the current to be very much retarded by fallen leaves, twigs, branches, and trees.

From this and former observations and lessons we learn, *that forests feed springs, brooks, and creeks, and that they tend to regulate the water supply for our large rivers and lakes. The difference between high and low water would not be so great as it is now if our whole country were well wooded.*

We also observed that washouts are not so common in the woods as they are in open fields. On many mountain slopes the soil would be entirely washed away if it were not held by roots of trees and other plants; *steep slopes should therefore never be deforested.* The roots of trees penetrate the ground to a much greater depth than the roots of grasses and herbs. Decayed roots, trunks, branches, and leaves have formed a layer of black soil in our deciduous forests several feet thick, much thicker than on the prairies. We have also observed that woods and groves are very important wind-breaks against the scorching south wind as well as against the icy blizzard.

§ 75. Forests and Man.

The teacher may here call the attention of the children to the endless variety of products which man derives directly and indirectly from the forests, and to some of the trades and occupations that depend on the forest.

Important as the influence of forests is upon the material welfare of man, its influence upon the mind of man must not be overlooked. To the first settlers on our shores the forests were the abode of thousands of warlike Indians. To all settlers in new countries, and to primitive man in general, the forests were always fraught with unknown and invisible danger, and the imagination of our children still peoples the big woods with hosts of bears, wolves, and all kinds of monsters. The dragons and other chimera which were killed by the knights of yore invariably lived in caves in the dark forest. For these obvious reasons settlers in a wild country always fought the forest with axe and fire. To-day we have conquered our forests. Groves and groups of shade trees shelter our farms and homes; they have inspired our poets, and they enable us to find rest and solitude whenever we are in need of them. From a nation of forest destroyers we must now become a nation of forest planters. Let us hope, and let us do our best, that we may accomplish the second task as thoroughly as we accomplished the first.

REFERENCES

Shaler. Chapter on Forests and Man in his Aspects of the Earth.
United States Department of Agriculture. The Relation of Forests to Farms. Tree Planting in the Western Plains. Forestry for Farmers.
References given under § 43, II.
Poems and Places, North America, Edited by Longfellow.
Guyot. The Earth and Man.

XIII

LAKE AND RIVER IN WINTER. FEBRUARY AND MARCH

Although in winter no such abundance of life is seen about us and in lakes and rivers as we found in the summer time, nevertheless there is plenty for our study and amusement. We can see fish swim under the ice, water plants cover the bottom, and aquatic insects are occasionally observed. Frogs and turtles have concealed themselves in the mud, and muskrats have retired to their domes, which we can now examine at our leisure. Many fishes can be caught through the ice, while others are seldom, if ever, seen during winter.

The next three paragraphs can as well be taken up during spring, summer, or fall; they are placed here on account of probable lack of time during the three seasons mentioned.

§ 76. The Common Sunfish, Pumpkin Seed. *Eupomotis gibbosus*

MATERIAL: A live sunfish in some convenient vessel; a freshly killed specimen or one preserved in alcohol. Of the latter the color must be carefully noted before the fish is put into alcohol, because alcohol changes the color of plants and animals. A few large minnows illustrate the structure of the fish almost as well. If the description given does not fit your sunfish, suit your description to your fish. Outdoor observations.

Observations. — How streams and all running water cut into the ground, bowlders of different colors, texture, and size; the rocks exposed in your vicinity, waterfalls, hard and soft coal.

The average size of this pretty fish in length and width is about that of a boy's hand, but you will catch many that are smaller. Can you see why it is called pumpkin seed? The color above is greenish with blue spots. The lower side and the cheeks are orange and the latter are marked with blue lines. It is a common fish in nearly all our lakes and rivers.

The sunfish, like all fishes, is well adapted to a life in the water, in which alone it is able to breathe. Watch it in the water and see how it seems to swallow some every second. This water passes out under two covers behind and below the eyes. Under these covers you find delicate, blood-red organs which look somewhat like small feathers and are attached to semicircular bones. These red organs are the gills, and with them the fish separates the air from the water and breathes the air. *The gills are the lungs of the fish, but it can breathe with them in water only.* When it is taken out of the water, it gasps for air, its gills stick together, and in a short time the fish dies of suffocation. A fish will exhaust the air from a pan of water in a short time, and must then be given fresh water. In the summer you must change the water for a fish more frequently than in the winter, because warm water holds less air than cold water. You can keep a few minnows in a tin pail for a long time. Feed them with a few crumbs of wheat bread three times a week and change the water after feeding.

The body of a fish is so shaped that it can easily cut through water. The sunfish has one pair of fins behind the gill covers and another pair on the abdomen. With these two pairs of fins, together with the large tail fin, the fish propels itself rapidly through the water, at the same time using the tail as a rudder. One fin on the back and another

Observations. — Fossils in the rocks or in collections.

below, in front of the tail, also aid the fish in keeping its direction and position, both acting like a keel-board on a boat. Does a boat in any way resemble the shape of a fish ?

The body of our little fish is covered with scales which overlap like shingles on a roof. Some fishes, as the bullheads and eels, have no scales, but all fishes are made slippery by a thin coating of slime. *The smooth covering of fishes is also of great advantage to them for moving rapidly in the water.* Sailors have found that a vessel whose submerged part is overgrown with seaweeds cannot make as good time as one whose hulk is smooth and clean.

The eyes of our sunfish are large and it does not take them long to see the worm when they are hungry. Their lips are organs of touch; the nasal openings do not go through into the mouth. Although they have no visible ears, they can perceive a shock caused by dropping something into the water. The tongue of most fishes is small, and is more an organ of touch than of taste, because all fishes swallow their food whole. The sunfish has a great many small teeth, but they only serve to seize and hold their prey.

All fish lay their eggs, called roe, in the water, and but very few of them pay any attention to the eggs or to the young. Indeed, the large ones generally eat all the eggs and the young which they find and catch. Under those conditions it is natural that but few eggs hatch and that few young fishes grow up; therefore most fishes lay an enormous number of eggs.

The sunfish lives on insects, worms, small minnows, and other small aquatic animals. When you catch a fish which is too small to be cleaned and eaten, you must carefully take it off the hook and put it back into the water. To let the small fish die would be cruel, and would be foolish, too, because it destroys your own sport.

Can you tell why pickerel and bass do not often eat a sunfish ?

§ 77. The Pickerel. . *Lucius lucius.*

MATERIAL: Similar to that for the sunfish. See the report on the fishes of your state.

Study the structure and color of the pickerel as you did those of the sunfish. Other fishes may be substituted for both of them.

The pickerel is the tyrant of our lakes and rivers. He attains a length of four feet and will swallow any fish that is not too big for that performance. Pickerel increase and grow very rapidly; one hundred and fifty thousand eggs were found in one. It is fortunate for other fish that most young pickerel are swallowed by larger members of their own kind.

§ 78. The Crayfish, or Fresh-water Crab.

MATERIAL: Live crayfish or alcoholic specimens. Observe in the classroom how a crab eats, walks, and swims ; observations at the lake or stream. Remove the large thorax carapace from an alcoholic specimen and show the branched gills below. Crayfishes must be procured in the fall if this lesson is given in winter.

The crayfish, or fresh-water crab, is found in nearly all the streams, lakes, and ponds of the United States and Canada, but it is probably absent from waters which are very soft, because such waters contain so little lime that the crayfish could not form its hard, outside skeleton, or crust, in them, as for this process lime is necessary.

Description. — The most conspicuous peculiarity of the crayfish is the hard crust, in which its body and limbs are encased.

The head and the chest are covered by a continuous shield, called the carapace. Behind the thorax we find six well-marked segments, which constitute the abdomen. The tail, which consists of five flaps, is attached to the last abdominal segment.

A crayfish has five pairs of legs. Observation of a live crayfish will teach you that the first pair, which are modified into the dreaded claws or pinchers, are not used for walking, but are offensive and defensive weapons. You can often see crayfishes walking slowly over the bottom; when, however, they wish to escape from danger, they strike the water downward with their tail and swim rapidly backward. The mouth of the crayfish is a rather large opening under the head; it is large enough to admit the head of a large pin, and is surrounded with very complicated prehensile and masticating organs.

The eyes of the crayfish are black and set on little stalks; they can be moved in all directions, and, when danger threatens, they are laid down in a furrow under the point of the head shield. The head is also provided with a pair of long, thread-like feelers, or antennæ. These long antennæ are delicate organs of touch. Just above them is a pair of shorter antennæ, which contain the organs of hearing. It is also highly probable that crayfishes can smell and taste, but the organs for these senses have not been made out with absolute certainty.

The color of the crayfish is a dark olive-green, which blends very nicely with the weeds and stones among which it lives. Crayfish very often live in holes, which they dig near the water. If the water dries up temporarily, they dig their holes deeper, until they reach moisture. If you remove the carapace of a crayfish, you will find the feather-shaped gills below it. By means of these gills the animal can breathe in air as well as in water, as long as its gills are kept moist. Transporting live crayfish is accomplished much better by putting them in wet grass or leaves than by keeping them in a pailful of water. You can induce the crayfish to leave their holes by stamping around them or by pouring water into them.

The food of the crayfish consists of snails, insect larvæ, tadpoles, small frogs, dead animals, and aquatic plants. Crabs should be kept in an aquarium by themselves, because they are likely either to eat or to injure other small aquarian animals.

During the winter, crayfish hide themselves. They reappear in early spring, and at this time the females are laden with eggs, which they carry beneath the tail, and which look like a mass of small berries. The eggs hatch in May or June, and the young, which closely resemble the adults, attach themselves to the mother for several days.

After a crayfish has been boiled, its calcareous shell looks red, and can now be easily separated from the white flesh within it. Vertebrate animals have their bones surrounded by flesh, but you see that the crayfish has its flesh surrounded by bones. The flesh and blood of the crayfish are white. How do you think this animal can grow in its tightly fitting armor? You can easily convince yourself that it will not stretch. Once or several times a year, the crayfish casts off its old integument. It splits below, and the animal crawls out of it, and pulls its legs, feelers, and eyes out too. Frequently you will find the crust, which looks just like the animal in shape, but is of a lighter color. It is not rare that one or more of the legs are broken in the process of moulting; but this does not seriously inconvenience our armored knight, as he has the peculiar power to repair any limb which has become detached or mutilated.

We must place the crayfish among the useful animals, because it acts as a scavenger in the water, and furnishes us with palatable food.

Teachers and pupils who would like to study aquatic animals are referred to the very interesting book, Fresh Water Aquaria, by G. C. Bateman. Though intended for England, it serves very well for this country.

§ 79. Review and Summary.

Fish, crabs, and other animals which breathe by means
of gills are compelled to live in the water, and most of
them can breathe only the air that is dissolved in water.
Some, like the trout, prefer cold streams; others, like the
tadpoles, like warm ponds.   For their food they depend
upon plants, upon smaller animals, or upon one another.
Fishes must lay a large number of eggs, because they do not
take care of them, and they would soon become extinct if it
were not for the enormous number of eggs produced.

Fresh-water and marine fish and salt-water lobsters furnish food and work for many thousands of people.   Our
lakes, rivers, ocean bays and banks, at one time, teemed
with fish, but ruthless slaughter has caused many good
grounds to be "fished out."   Several states, as well as the
Federal Government, have established fish-hatcheries for
the purpose of stocking, or restocking, our waters with
valuable food and game fish.   The fish in lakes and rivers
furnish so much healthful recreation to young and old, to
rich and poor, that their preservation is of even greater
importance than the preservation of other animals.

*Catch no fish in their spawning season, and promptly return
to the water all fish that are too small to be used, is the rule
of every intelligent fisherman.*

Much damage is often done to river fish by city and farm
sewerage and by the waste product of factories.

XIV

REVIEW OF ANIMAL LIFE

§ **80**. *a. Classes of animals.* — The lowest animal which we have studied is the Earthworm. Other well-known worms are the Roundworm (*Ascaris lumbricoides*) and the Tapeworm, both parasitic in the intestines of man. Common aquatic forms are the Leeches, generally called Bloodsuckers by children.

All worms have a soft, elongated, and contractile body. The body consists of rings or segments, and there is no distinct head, thorax, or abdomen. Their body is not protected against the evaporation of moisture, and therefore they can live in moist places only. When do earthworms move about? What becomes of them if they are accidentally placed on dry boards or stones? Can they creep over dry ground?

You must remember that the maggots of flies, the grubs of beetles, and the larvæ of butterflies and moths, although often called worms, are not worms.

Articulates. — These animals, to which the *Crabs, Spiders,* and *Insects* belong, are much more highly organized than the worms. The covering of their body is hard, and does not restrict them to moist places. This hard armor consists of several rings or segments, and does, therefore, not seriously interfere with their motions. All of them show more or less distinctly a head, a thorax, and an abdomen. They have paired legs, and are provided with distinct organs of sight and touch, and many have also organs of scent and

302

hearing.   The highest animals among the articulates are the insects.

The *Molluscs* form another class, with which you are quite familiar.   The oyster, the clams, the different kinds of snails and slugs, belong to this class.   Most of them live in hard calcareous shells, which protect their soft bodies.

Worms, articulates, and molluscs have no bony skeleton. Fishes, frogs, snakes, birds, and mammals do have a bony skeleton, whose principal part is the backbone, or vertebral column, which consists of a large number of bones called vertebræ.

Worms, articulates, and molluscs belong to the great division of invertebrates ; fishes, frogs, snakes, birds, and mammals are vertebrates.

The lowest among the vertebrates are the *Fishes*.   The structure of their breathing organs, the gills, restricts them to a life in the water.   Fins are their limbs of locomotion, their sense organs are not highly developed, and their brain is very small.

The next higher class, the *Amphibians*, comprises such animals as frogs, toads, and salamanders.   These animals lay their eggs in water like fishes, but the young do not, at first, resemble the adult, and pass through a metamorphosis.   During the polliwog state they can breathe only in water by means of gills.   After some time their lungs develop; they can now breathe in the air, and they acquire the form of the adults.   Amphibians are covered with a soft, naked skin; they live in or near water, and in moist, shady places.

The next higher class, the *Reptiles*, deposit their eggs in the ground, where they are hatched by the warmth of the sun.   The young breathe at once by means of lungs, and resemble the adults.   To this class belong snakes, lizards. turtles, and alligators.   Reptiles are covered with scales or with horny plates. The smaller ones are well adapted to live in holes or in rock crevices. The blood of fishes, amphibians, and reptiles has about the temperature of the air or the water in which they live, and they always feel cool to our touch, and for this reason they are often called *cold-blooded animals.*

Higher than the reptiles stand the *Birds* and the *Mammals.* The blood of these two classes is kept at a uniformly high temperature winter and summer; they are therefore often spoken of as *warm-blooded animals.* For reasons given under "Review of Birds and Review of Mammals," the latter must be placed at the head of the animal kingdom.

This brief sketch is not even an attempt at a complete classification; in fact, many important classes of marine animals have not been mentioned at all.

b. Conditions for animal life. — Animals depend *for food* either upon plants or upon other animals. *All of them need air,* which they breathe directly by means of lungs, or absorb it from the water by means of gills. *Nor can one of them live entirely without water,* although some find a sufficient amount of it in the food which they eat. *Warmth and light,* as you can easily prove, are also important factors in animal life.

XV

THE EFFECT OF WATER AND ICE, OF ANIMALS AND PLANTS, AND OF HEAT UPON THE EARTH

MATERIAL: Pieces of rock containing fossils, coral, slate, and coal, with impressions of plants; peat; pictures of a volcano; pieces of lava. If practicable, take the children to a place where they can collect and observe fossils in their natural position in the rocks. About localities for collecting fossils consult the Geological Reports of your state, write to the nearest high school or to your state geologist. Review briefly the chapter on "Geological Effect of Water."

§ **81.** We have learned by former observations that running water erodes the rocks and carries the pulverized material into lakes and river valleys and finally into the ocean. A great deal of the soil in the Mississippi basin and on the Atlantic slope has been made in this way. Many large rivers of the world have formed fertile plains and deltas by means of the soil they brought down from the highlands near their sources. This work has been going on since the rivers began to flow, and it is still going on, because we know that all deltas are growing seaward. Many rivers form no deltas, because the material they carry into the sea is taken up by ocean currents and swept into the deep sea, where its accumulation is not observed. What we have said about the transportation of fine soil you can all understand and prove by observation, but it is more

Note. — Arrange with your pupils to make such observations as are calculated to carry on the work they have begun.

difficult to prove where the detached stones or bowlders came from.

Everywhere in our Northern States, but especially along rivers, streams, and lake shores, we find large and small stones which do not consist of the same material as the nearest stratified rock; they are nearly always much harder than the nearest sandstone and limestone, and show a great variety of color and texture. Occasionally a bowlder is found which is so large that twenty horses could not move it an inch. Shall we believe that once water rushed over all our Northern States with such terrific velocity that it could move and scatter these monstrous rocks? No, we must look to another agency for an explanation of the presence of these bowlders.

Geologists have proved beyond a reasonable doubt that there was a time when this country was covered by a sheet of ice thousands of feet thick. This ice began to form in Canada and in the Great Lake regions and was thickest there. On account of the slope of the land, and by its own weight, it crept slowly southward until it covered all our Northern States from Long Island to near the mouth of the Ohio River; from there its southern margin extended in a north-westerly direction to the mouth of the Missouri, and extended up that river into the Yellowstone Park region. This slowly moving ice field broke many pieces of rock from mountain and hill sides, and others it carried along on the bottom. When the climate grew warmer again, the ice melted and dropped stones and soil where we find them to-day. The bowlders are of the same structure as the rocks farther north over which the glacier flowed. The farther south one goes, the smaller the bowlders grow, and finally none are found any more. Some of you may know

---

They should write up, in brief form, the results of the trips and walks they take from time to time.

that much copper is mined on the south shore of Lake Superior, especially on the Keweenaw Peninsula. Large pieces of almost pure copper are common here, and have no doubt been formed where we find them to-day. If, however, you should now and then find large pieces of metallic copper in the glacial soil at St. Paul and Minneapolis, and even as far south as Illinois and Iowa, you would find this metal far from its natural source or origin, and as a thinking person you would ask, How did it come to be there? As a matter of fact, such pieces of copper are frequently found in the regions mentioned, and generally show that they have been pressed or rolled by some heavy mass. The only reasonable explanation is that they were dropped by the melting ice sheet. Geologists have calculated from receding river gorges, like that of Niagara below its falls, and that of the Mississippi between Fort Snelling and Minneapolis, that the ice probably disappeared between ten and twenty thousand years ago; but nothing is known about the length of the glacial epoch itself. There is good evidence that northern Europe had its glacial epoch as well as North America.

For information about the probable causes of the glacial epoch and for further proofs of it, the teacher is referred to some text-book on geology. On the recession of the Mississippi gorge, see Geology of Hennepin County in the Final Report of the Minnesota State Geologist.

§ **82**. *Formations consisting of vegetable débris.* — When we studied the mosses, we learned that thick layers of peat have been formed, and are still being formed, by the decay of mosses and other vegetation. *The coal beds, which are from less than an inch to forty feet thick, were also formed by plants.* This is proved by the many impressions of plants we find in them, and also by roots and tree-trunks which have been discovered in them. The coal plants were mostly tree-like ferns, horsetails, and lycopods, which grew in vast swamps. Here one generation after the other fell into the shallow water, the land was slowly sinking, the salt water of the ocean killed the plants, and the thick mass of *débris* was covered by clay and sand, which was carried in from the adjacent land. A slow process of decay, and the great pressure exerted upon them, has left nothing of these grand forests but the coal, which was in their trunks and foliage. Geologists believe that millions of years have passed since the coal plants flourished.

Formations made by the remains of animals. — Let us carefully examine these pieces of limestone and marble. We find that they consist entirely of shells and other calcareous remains of animals. We also know that corals form islands and reefs, the latter sometimes being hundreds of miles long. Rock formed by animal remains is limestone; and it has been calculated that the average thickness of this rock over all the continents is about five hundred feet. The white chalk cliffs of England and France are also the calcareous remains of very small animals.

§ **83.** *Finally, we must mention the interior heat of the earth as an important factor in determining the character of its surface.* Whether the earth's interior is solid or liquid is still a mooted question; but the high temperature in deep mines, the boiling waters ejected by geysers, and the glowing lava, or melted rock, and steam erupted by volcanoes, is sufficient evidence that a temperature high enough to melt all the rocks, metals, and minerals we know of prevails in the interior of the earth. Volcanoes often eject enough lava and ashes to cover many square miles, and have buried whole cities. Hundreds of square miles in the Rocky Mountains are covered by lava, which very long ago came up through great fissures in the stratified rocks. It is a generally accepted theory that the whole earth was once a glowing liquid mass, like melted iron. As this heated globe continually gave off large amounts of heat into space, a solid crust was formed around the melted interior. When a body cools, it contracts. The contraction of the globe caused its solid crust to be thrown up into huge wrinkles, which are now the mountain ranges; while the lower places were filled up by the waters of the ocean. Since that time, many shallow seas have been partially filled up by deposits carried into them by rivers and waves, and have later been lifted up, and become dry land. Over three-fourths of the surface rock in North America is stratified, and contains marine fossils, which is positive proof that the ocean once covered all of this area. At the present time, layers of ocean-made rock form the very peaks of many snow-clad mountains. In other places, land, once formed, sank again; and forests, fields, and towns were buried beneath the waves. Changes like the ones just described have been going on ever since there was land and water on the globe, and they are still going on; but with the exception of those taking place on the shores of the ocean, and those resulting from severe earthquakes and volcanic eruptions, they take place so slowly that only trained observers notice them. Do rivers raise or lower the land? Do our lakes become deeper, or can you detect agencies that tend to fill them in? The teacher should point out how streams and rivers have carved, or eroded, soil and rock.

REFERENCES

Tarr.   Physical Geography.   The Macmillan Company.
Tarr.   Elementary Geography.   The Macmillan Company.
Heilprin. The Earth and its Story.   Silver, Burdett & Co.

XVI

ANIMATE AND INANIMATE NATURE; PLANTS, ANIMALS, AND MAN IN THEIR RELATION TO ONE ANOTHER

§ **84.** *The earth is acted upon by the physical forces of nature, and has been thus influenced ever since it came into existence.* Earthquakes, volcanic eruptions, ocean waves and currents, rivers, rains, and winds destroy land in one place and make new land in other places. *Compared with the work of these agencies, the influence of plants and animals seems insignificant, although it is of great importance for man.*

Plants increase the fertility of the soil by the decay of their roots, leaves, stalks, and wood. The dark color in the surface soil is due to the finely divided particles of decayed vegetation. In forest regions this black surface soil is sometimes three feet thick, while in the prairie regions it is seldom more than a foot thick. This admixture of decayed vegetation is of the greatest importance to the agriculturist, because very few plants grow in pure sand or in pure clay. The roots of plants penetrate into the crevices of rocks, and, by their growth, widen the cracks; and the carbonic acid, which is one of the products of the decay of plant and animal tissue, accelerates the rotting of rocks. Rocks which are not below the frost line are much changed by that agency. The moisture, penetrating the fine vertical cracks and accumulating under the horizontal layers, expands very powerfully when it freezes, and thus breaks the rock into many small pieces, and prepares the way for roots.

Still less conspicuous than the influence of plants is that of animals. Worms, ants, gophers, and other burrowing animals really subject the soil to a slow process of cultivation.

In northern countries, plants have formed extensive peat bogs, and large accumulations of vegetable *débris* are forming in the Everglades of Florida and in other sub-tropical and tropical swamps, and even the coal beds are the remains of decayed forests. In the making of rocks, however, animals have played a much more important part than plants. Chalk, limestone, and marble are nothing but the calcareous remains of animals.

Plants in their relation to inanimate nature and to animals. Plants need air and water and several mineral substances for the building up of their tissue. They also depend on warmth, light, and winds for favorable conditions of life. Many plants depend upon animals for cross-fertilization or for the dissemination of their seeds.

Animals also depend upon plants and upon inanimate nature for the conditions of their existence. They need water, air, warmth, and light as well as plants, although light is not so all-important for them as it is for plants. Animals are not able to feed upon the lifeless mineral matter in the soil. This power is possessed by plants only, and animals depend on them for food. No animals could exist if there were no vegetation on the earth.

When plants and animals decay, a part of the substance of their bodies is returned to the atmosphere in the form of gases; the mineral matter and the water are returned to the soil, whence they originally came.

§ **85.** *Man himself is connected in many ways with animate and inanimate nature.* He cannot exist without air, water, and light; climate and temperature influence his well-being, and he depends for food on the soil he tills.

Man has, however, on his part, exerted a most powerful influence upon nature. By converting forests and prairies into fields, by draining swamps, by building cities, roads, and railways, he has changed the aspect of whole continents. He has exterminated many animals and domesticated others, causing them to increase far beyond the number which they could have reached in a wild state. Man's physical strength is not great, if we compare him with the largest animals; but his hand, directed by a superior intellect, is able either to subdue or exterminate the fiercest and largest animals. His intellect has invented means which enable him to cross the never-resting ocean, as well as the dreary desert; he can live under the heat of an equatorial sun and on the ice fields of Greenland; the sunlight paints his pictures, and the lightning's flash conveys his thoughts around the globe; he is fulfilling the prophetic words, "Replenish the earth and subdue it, and have dominion over the fish of the sea and over the fowl of the air, and over everything that moveth upon the earth."

Man is the only being on earth that is conscious of the laws and workings of nature.

§ **86**. Just as every plant, every animal, and every stone on the earth is only a part of the whole and is intimately connected with the whole, so the earth itself is only one of the large heavenly bodies of our solar system. *For its life-giving warmth and light, the earth depends on the sun;* and if his blazing light should ever be extinguished, all life on our earth would become extinct.

APPENDIX

I. FIELD WORK AND FIELD LESSONS

WHILE you are studying in school the topics treated in a certain chapter, let the pupils make outdoor observations for the topics of the next chapter. You will find many suggestions for these observations in the foot-notes, but you will obtain the best results if you carefully adapt the Nature Study work to the locality and conditions of your school.

Besides the observations, which the pupils should make for themselves, several well-planned field lessons should be given by the teacher, who must be thoroughly familiar with the woods, or field, or valley where the lesson is to be given. It is a good plan to tell the children what they are expected to look for and to find. About the time, the distance, and the number of pupils for these field lessons, the teacher is the best judge. This book treats the subject of Nature Study without taking into account the summer vacation. As summer is the best time for observing plant and animal life, the teacher should give the pupils a syllabus of observations to be made during vacation, and should ask for reports on these at the next opening of school.

II. MATERIAL

It is not expected that every teacher will be able to procure all the material mentioned for each lesson. The pupils will gladly collect anything they can find; but the teacher should observe what is said in the preface on this subject, or the whole subject of Nature Study may fall into disrepute. The teacher who would undertake to teach Nature Study without outdoor observations, field lessons, and material will simply make a farce out of it, and increase the evil of bookishness and verbalism; in other words, she will lead the children to talk about things that they do not know.

III. HOW TO PRESERVE PLANTS

Lay the plants or twigs between sheets of newspaper, cover the paper with a fairly smooth board, place stones or other objects weighing from five to twenty pounds on the board. The first paper with the plants may be laid on a dry wooden floor or table; use as much paper and as many boards as you may require. Change the paper every twenty-four hours, until the plants feel entirely dry to the touch, then fasten them on uniform sheets of white paper by means of gummed paper strips. Write the name of the plant, the date of collection, and other desirable data on each sheet. Fruit and seed of plants are best kept in paper boxes. Keep these plants in a perfectly dry place.

Some care should be exercised that children do not come in contact with Poison Ivy, as this results sometimes in serious cases of skin poisoning. Study carefully the figures of the poison ivy, and of the Virginia creeper.

The Virginia Creeper is also called Woodbine and American Ivy. A very common, woody vine, climbing extensively by tendrils as well as by rootlets. Small greenish flowers in July; *black or bluish berries*, ripe in October, *resembling wild grapes*, and remaining on the vines over winter. *Each leaf generally consisting of five leaflets;* leaves turning bright crimson in fall. Leaves and berries are both harmless.

The Poison Ivy, or Poison Oak. *Leaves consisting of three leaflets each; berries whitish*, remaining on the plant over winter. In central Minnesota this plant is generally a low, erect shrub, from six to eighteen inches high, but it sometimes climbs over rocks or ascends trees. Some people are seriously poisoned by touching the plant.

Children cannot be too strongly warned against eating plants or fruits of whose harmless nature they are not absolutely certain. A child that has eaten of a poisonous plant should at once be given something that will cause vomiting, and a physician should be called as soon as possible.

See *V. K. Chestnut.* Some Common Poisonous Plants.

F. V. Coville. Recent Cases of Mushroom Poisoning. Both published by the United States Department of Agriculture.

FIG. 59.— VIRGINIA CREEPER. *Ampelópsis quinquefolia.*
Much reduced.

Fig. 60. — Poison Ivy, or Poison Oak. *Rhus toxicodendron.*
Much reduced.

IV. PRESERVATION OF INSECTS

Place about one hundred grains of potassium cyanide, broken up into small pieces in a large-mouthed bottle, add water to the depth of half an inch, then add plaster of Paris until a dry cake is formed. Wipe out with a cloth the dry plaster adhering to the sides of the bottle. After the bottle has stood loosely corked for an hour, it is ready for use.

Any insects placed in this poison bottle will die at once. They may be pinned in cigar boxes, or may be preserved in paper envelopes for future use. Always keep your poison bottle tightly corked, the cyanide fumes are very poisonous. Children should not prepare the poison bottle, but they can safely use it. In order to make the legs and wings of dried insects again pliable, place the insects in a tin pail, put a piece of moist blotting paper in it, and keep the pail closed for about half a day.

V. AQUARIUMS

Common butter jars, glass fruit jars, and electric battery jars, and almost any kind of stone and glass vessels are suitable for small aquariums. Place a few stones, pebbles, and some clean gravel on the bottom. Add a few small water plants, such as the little floating duckweeds, tufts of the green threadlike algæ, bits of water-pest, etc. Do not place too many plants nor too many animals in one aquarium. Experience must teach you what animals can be placed in the same aquarium without eating one another. Minnows, crabs, snails, all kinds of aquatic insects, tadpoles, small water plants, can be kept in an aquarium. Minnows and tadpoles eat crumbs of wheat bread or goldfish food; crabs, water beetles, and a few other insects will eat earthworms, bits of raw meat, flies, etc. Some animals will feed on the minute plants and animals in the aquarium. Feed your animals every other day in summer and every third or fourth day in winter, but give them no more than they will eat. Let little or no direct sunlight fall into the aquarium, keep the water fresh, avoid too great changes of temperature when you change the water. Aquariums with winged insects must be covered with gauze to prevent the insects from flying out.

See *Miall*. Aquatic Insects. The Macmillan Company.

VI.　OTHER COLLECTIONS

Children must not make collections of birds and birds' eggs. A small collection of mounted birds might be valuable for the school, but the children must not do the collecting, nor should they be allowed to collect eggs.

Bones of mammals should be boiled, after which the flesh can be picked off, but they should be prepared by the teacher. The teacher can also procure good material at the meat market. A clean bone or a clean, fresh piece of meat are not objectionable. The nests of birds may be taken after the young birds have left them. Such nests should be disinfected by being placed in gasolene for about fifteen minutes. Do not place the gasolene near a lighted lamp or near fire of any kind.

VII.　ALCOHOLIC MATERIAL

Frogs, tadpoles, beetles, and other insects, as well as fruits and roots of plants, can be preserved for an indefinite time in seventy per cent alcohol; but alcoholic material loses most of its natural color.

VIII.　SOME HELPFUL LITERATURE

1. Write to the Agricultural Experiment Station of your state for a list of its publications.

2. Write to the Secretary of Agriculture, Washington, D. C., for a list of publications issued by the United States Department of Agriculture.

3. Below is a brief list of books that will furnish interesting reading for somewhat mature persons.

Any of the books of *John Burroughs*.

Any of the books of *C. C. Abbott*.

Ingersoll.　Wild Neighbors.

Cornish.　Animals at Work and at Play.

Rogers.　Hunting American Big Game.

Edwards.　Camp Fires of a Naturalist.

Thompson.　The Boys' Book of Sports. An excellent book for young and old to have with you when out camping.

Arthur and MacDougal. Living Plants and Their Properties. $1.25. University Book Store, Minneapolis, Minn.

MacDougal. Physiological Botany.

4. See the books mentioned in the text.

It is entirely wrong to begin the study of nature with books. First observe the life about you, then you are ready for special and for general reading.

5. FARMERS' BULLETINS.

These bulletins are sent free of charge to any address upon application to the Secretary of Agriculture, Washington, D. C.

[Only the bulletins named below are available for distribution.]

No.

15. Some Destructive Potato Diseases: What They Are and How to Prevent Them.
16. Leguminous Plants for Green Manuring and for Feeding.
18. Forage Plants for the South.
19. *Important Insecticides:* Directions for Their Preparation and Use.
20. *Washed Soils:* How to Prevent and Reclaim Them.
21. Barnyard Manure.
22. Feeding Farm Animals.
23. Foods: Nutritive Value and Cost.
24. Hog Cholera and Swine Plague.
25. Peanuts: Culture and Uses.
26. Sweet Potatoes: Culture and Uses.
27. Flax for Seed and Fibre.
28. *Weeds; and How to Kill Them.*
29. Souring of Milk, and Other Changes in Milk Products.
30. Grape Diseases on the Pacific Coast.
31. Alfalfa, or Lucern.
32. Silos and Silage.
33. Peach Growing for Market.
34. Meats: Composition and Cooking.
35. Potato Culture.
36. Cotton Seed and Its Products.
37. Kafir Corn: Characteristics, Culture, and Uses.

38. Spraying for Fruit Diseases.
39. Onion Culture.
40. Farm Drainage.
41. *Fowls: Care and Feeding.*
42. Facts About Milk.
43. Sewage Disposal on the Farm.
44. Commercial Fertilizers.
45. Some Insects Injurious to Stored Grain.
46. Irrigation in Humid Climates.
47. Insects Affecting the Cotton Plant.
48. The Manuring of Cotton.
49. Sheep Feeding.
50. Sorghum as a Forage Crop.
51. *Standard Varieties of Chickens.*
52. The Sugar Beet.
53. *How to Grow Mushrooms.*
54. *Some Common Birds in Their Relation to Agriculture.*
55. The Dairy Herd : Its Formation and Management.
56. Experiment Station Work—I.
57. Butter Making on the Farm.
58. The Soy Bean as a Forage Crop.
59. *Bee Keeping.*
60. Methods of Curing Tobacco.
61. Asparagus Culture.
62. Marketing Farm Produce.
63. Care of Milk on the Farm.
64. *Ducks and Geese.*
65. Experiment Station Work   II.
66. *Meadows and Pastures.*
67. Forestry for Farmers.
69. Experiment Station Work — III.
70. The Principal Insect Enemies of the Grape.
72. Cattle Ranges of the Southwest.
73. Experiment Station Work — IV.

6. Write to Cornell University Agricultural College, Ithaca, N. Y., for a list of Bulletins and Teachers' Leaflets.

7. List of Agricultural Experiment Stations in the United States and Canada

(Address mail to them in the following manner: Agricultural Experiment Station, St. Anthony Park, Minn.)

UNITED STATES

State	Post-Office	State	Post-Office
Alabama (College) .	Auburn.	Missouri . .	Columbia.
Alabama (Canebrake) . .	Uniontown.	Montana . .	Bozeman.
		Nebraska . .	Lincoln.
Arizona	Tucson.	Nevada . . .	Reno.
Arkansas	Fayetteville.	New Hampshire . .	Durham.
California	Berkeley.	New Jersey (State) .	New Brunswick.
Colorado	Fort Collins.		
Connecticut (State).	New Haven.	New Jersey (College)	New Brunswick.
Connecticut (Storrs)	Storrs.		
Delaware . .	Newark.	New Mexico . . .	Mesilla Park.
Florida	Lake City.	New York (State) .	Geneva.
Georgia	Experiment.	New York (Cornell).	Ithaca.
Idaho	Moscow.	North Carolina . .	Raleigh.
Illinois	Urbana.	North Dakota . .	Fargo.
Indiana	Lafayette.	Ohio	Wooster.
Iowa	Ames.	Oklahoma	Stillwater.
Kansas	Manhattan.	Oregon	Corvallis.
Kentucky	Lexington.	Pennsylvania . . .	State College.
Louisiana (Sugar) .	New Orleans.	Rhode Island . . .	Kingston.
Louisiana (State) .	Baton Ronge.	South Carolina . .	Clemson College.
Louisiana (North) .	Calhoun.		
Maine	Orono.	South Dakota . .	Brookings.
Maryland . .	College Park.	Tennessee . .	Knoxville.
Massachusetts (State) . . .	Amherst.	Texas	College Station.
Massachusetts (Hatch) . . .	..	Utah	Logan.
		Vermont . .	Burlington.
Michigan . .	Agricultural College.	Virginia . .	Blacksburg.
		Washington . .	Pullman.
Minnesota . .	St. Anthony Park.	West Virginia . .	Morgantown.
		Wisconsin	Madison.
Mississippi . . .	Agricultural College.	Wyoming	Laramie.

List of Agricultural Experiment Stations. *Continued*

CANADA

Province	Post-Office	Province	Post-Office
Manitoba . .	Brandon.	Nova Scotia . .	Nappan.
N.W. Territory . .	Indian Head.	Ontario	Ottawa.
British Columbia .	Agassiz.		

INDEX

[The foot-notes and publications mentioned in the text are not indexed. Binomials are indexed under both words.]

A

About home. March to June, 1.
Acer dasycarpum, 130.
Acer saccharium, 130.
Adiantum pedatum, 239.
Agricultural experiment stations, list of, in the United States and Canada, 321, 322.
Agriculture, influence of, upon man, 228.
Agriculture, stock-raising and, 227.
Alcoholic material, 318.
Ambrosia artemisiæfolia, 58.
Ambrosia trifida, 58.
American big game, 278.
American goldfinch, 162.
Ampelopsis quinquefolia, 315.
Animal life, conditions for, 304.
Animal life in the woods, 154.
Animal life, review of, 302.
Animals, classes of, 302.
Animals, domestic, 94.
Animals, influence of man upon, 119.
Animals in their relation to plants and to inanimate nature, 311
Animals, our duty to, 121.
Animals, the effect of, upon the earth, 308.
Anosia plexippus, 202.
Anthemis coluta, 58.
Anthers, 16.
Aphids, 8.
Apple tree, 4.
Aquariums, 317.

Arctium Lappa, 58.
Asclepias Cornuti, 195.
Ash-leaved maple, 70.
Ash, white, 131, 139.
Aspen, 137.
Asp, quaking, 127.
August, prairie flowers in, 50.
Autumn, the woods in, 238.
Avena fatua, 192.

B

Balm of Gilead, 129.
Balsam poplar, 129.
Baltimore oriole, 158.
Bank, sand, 40.
Bare-soil moss, 245.
Barley, 183.
Basswood, 135.
Bat, 76.
Bear, 285.
Bee, 204.
Beetle, potato, 212.
Begonias, 87.
Beneficial, insects, to man, 9.
Berry, sugar, 135.
Betula papyrifera, 131.
Big game, American, 278.
Birch, canoe, 131.
Birch, paper, 131, 142.
Birds, closing remarks on, 165.
Birds, resident in our Northern States, 262.
Birds, review of the, 271.
Birds seen in the field, 216.

Review of the birds, 271.
Review of the mammals, 286.
Rhus toxicodendron, 316.
Roadsides, 58.
Robin, 11.
Rocks, how made, 42.
Rock maple, 130.
Rodents, 177.
Roots, 90.
Rose-breasted grosbeak, 160.
Rose, garden, 12.
Russian thistle, 65.
Rust, 250.
Ruminants, 105.
Rye, 183.

S

Salsola Kali, 58.
Sand bank, 40.
Scarlet oak, 134, 149.
Sciurus Carolinensis, 171.
Sciurus Hudsonius, 171.
Sepals, 16.
Setaria glauca, 192.
Sheep, 100.
Silkweed, 195.
Silver maple, 130, 139.
Skunk, 274.
Smut, 250.
Soil, field and garden, 224.
Soft maple, 130, 139.
Sparrow, chipping, 162.
Sparrow, English, 73.
Sparrow, house, 73.
Spermophilus tridecemlineatus, 220.
Sphagnum cymbifolium, 245.
Spider, 80.
Spinus tristis, 162.
Spizella socialis, 162.
Spring, flowers of early, 153.
Spruces, 255.
Squirrel, flying, 171.
Squirrel, fox, 171.
Squirrel, gray, 171.
Squirrel, red, 171.
Squirrels, tree, 171.
Stamens, 16.
Stem, 90.

Stigmas, 16.
Stock-raising and agriculture, 227.
Striped gopher, 220.
Sugar berry, 135.
Sugar maple, 130, 144.
Summary of life in the woods, 289.
Summer foliage, the woods in their, 231.
Sunfish, 295.
Sunflower, 53.
Sunflower, false, 59.
Swallows, 27.
Swamp hickory, 134, 147.
Sweet clover, 64.

T

Tachina flies, 10.
Tamias striatus, 167.
Teachers' leaflets, 320.
Tent caterpillar, 6.
Terms, botanical, 15.
Thistle, Canada, 60.
Thistle, Russian, 65.
Thrasher, brown, 158.
Tilia Americana, 135.
Toadstools, 248.
Tree, apple, 4.
Tree squirrels, 171.
Trees and shrubs, the leaves of our, 231.
Trees, fruit, 4.
Trees in their winter condition, 127.
Trees, introduction, 123.
Trees, shrubs, and vines, the fruit of, 235.
Tulip, 1.
Tulipa Gesneriana, 1.
Tympanuchus Americanus, 218.
Tyrannus tyrannus, 217.

U

Ulmus Americana, 133.
Underbrush, 177.

V

Vegetables, garden, 3.
Vireo olivaceus, 165.

"AN IDEAL BOOK ON NATURE STUDY."

CITIZEN BIRD.

Scenes from Bird Life in Plain English for Beginners. By
MABEL OSGOOD WRIGHT and ELLIOTT COUES. With One
Hundred and Eleven Illustrations by Louis Agassiz Fuertes.
12mo, Cloth, $1.50, *net*.

This first issue of The Heart of Nature Series— *Citizen Bird*— is
in every way a remarkable book. It is the story of the Bird-People
told for the House-People, especially the *young* House-People, being
dedicated "To All Boys and Girls who Love Birds and Wish to Protect Them."

It is not a mere sympathetic plea for protection. It shows how Citizen Bird "works for his own living as well as ours, pays his rent and
taxes, and gives free concerts daily"; is scientifically accurate in description of anatomy, dress, and habits; and is illustrated by over one
hundred engravings in half tone, together with descriptive diagrams,
and has a valuable index of some one hundred and fifty-four American
birds.

It is a question when one becomes too old to enjoy such a delightful and entertaining book.

TOMMY-ANNE

AND

THE THREE HEARTS.

By MABEL OSGOOD WRIGHT. With many Illustrations by Albert
D. Blashfield. 12mo, Cloth, Colored Edges, $1.50.

This book is calculated to interest children in nature, and grown folks,
too, will find themselves catching the author's enthusiasm. As for Tommy-
Anne herself, she is bound to make friends wherever she is known. The
more of such books as these, the better for the children. One Tommy-
Anne is worth a whole shelf of the average juvenile literature." —*Critic*.

" Her book is altogether out of the commonplace. It will be immensely
entertaining to all children who have a touch of imagination, and it is
instructive and attractive to older readers as well." —*Outlook*.

The work is probably the most charming nature-book for children
published this year." —*Dial*.

THE MACMILLAN COMPANY,

66 FIFTH AVENUE, NEW YORK.

FIRST BOOK IN
PHYSICAL GEOGRAPHY.

By RALPH STOCKMAN TARR, B.S., F.G.S.A., Professor of Dynamic Geology and Physical Geography at Cornell University. 12mo, Half Leather, $1.10, *net.*

The striking success of Tarr's Elementary Physical Geography in high schools has led to the preparation of this *First Book*, which is designed for use in public and private schools requiring a somewhat shorter course than is given in the Elementary Physical Geography. Its claim to attention lies in its presentation of physical geography in its modern aspect. The main emphasis is laid upon physiography, and all the features that have contributed to the rapid introduction of the earlier books are retained in simpler form.

ELEMENTARY
PHYSICAL GEOGRAPHY.

By R. S. TARR. 12mo, Half Leather, $1.40, *net.*

The widespread and increasing use of Tarr's Elementary Physical Geography, due originally to the recent and general change in methods of teaching the subject, has received a renewed impetus during the present year from the enthusiastic commendations of the teachers in the public schools of Chicago, Brooklyn, Philadelphia, Kansas City, and many other important centres.

ELEMENTARY GEOLOGY.

By R. S. TARR. 12mo, Half Leather, $1.40, *net.*

This book, published in February, 1897, is now generally recognized as the most attractive and scientific presentation of the subject for high schools. Many important schools have already adopted it.

THE MACMILLAN COMPANY,
66 FIFTH AVENUE, NEW YORK.

BOOKS ON NATURE.

BADENOCH (L. N.). — **The Romance of the Insect World.**
By L. N. BADENOCH. With Illustrations by Margaret J. D.
Badenoch and others. *Second Edition.* Gilt top, $1.25.

"The volume is fascinating from beginning to end, and there are many
hints to be found in the wisdom and thrift shown by the smallest animal
creatures." — *Boston Times.*

'A splendid book to be put in the hands of any youth who may need
an incentive to interest in out-door life or the history of things around
him." — *Chicago Times.*

BRIGHTWEN. — **Inmates of My House and Garden.** By
Mrs. BRIGHTWEN. Illustrated. 12mo, $1.25.

"One of the most charming books of the season, both as to form and
substance." — *The Outlook.*

"The book fills a delightful place not occupied by any other book that
we have ever seen." — *Boston Home Journal.*

GAYE. — **The Great World's Farm.** Some Account of Nature's
Crops and How They are Grown. By SELINA GAYE.
With a Preface by G. S. Boulger, F.L.S., and numerous
Illustrations. 12mo, $1.50.

The University of California expressly commends this to its affiliated
secondary schools for supplementary reading.

"It is a thoroughly well-written and well-illustrated book, divested as
much as possible of technicalities, and is admirably adapted to giving young
people, for whom it was prepared, a readable account of plants and how
they live and grow." — *Public Opinion.*

"One of the most delightful semi-scientific books, which every one enjoys
reading and at once wishes to own. Such works present science in the
most fascinating and enticing way, and from a cursory glance at paragraphs
the reader is insensibly led on to chapters and thence to a thorough reading from cover to cover. . . . The work is especially well adapted for
school purposes in connection with the study of elementary natural science,
to which modern authorities are united in giving an early and important
place in the school curriculum." — *The Journal of Education.*

THE MACMILLAN COMPANY,

66 FIFTH AVENUE, NEW YORK.

HUTCHINSON. — **The Story of the Hills.** A Book about Mountains for General Readers and Supplementary Reading in Schools. By H. N. HUTCHINSON, author of "The Autobiography of the Earth," etc. Illustrated. $1.50.

"A book that has long been needed, one that gives a clear account of the geological formation of mountains, and their various methods of origin, in language so clear and untechnical that it will not confuse even the most unscientific." — *Boston Evening Transcript.*

"It is as interesting as a story, and full of the most instructive information, which is given in a style that every one can comprehend. . . ."
— *Journal of Education.*

INGERSOLL. — **Wild Neighbors.** A Book about Animals. By ERNEST INGERSOLL. Illustrated. 12mo, Cloth. $1.50.

JAPP (A. H.). — **Hours in My Garden,** and Other Nature-Sketches. With 138 Illustrations, $1.75.

"It is not a book to be described, but to be read in the spirit in which it is written — carefully and lovingly." — *Mail and Express.*

"It is a book to be read and enjoyed by both young and old."
— *Public Opinion.*

POTTS (W.). — **From a New England Hillside.** Notes from Underledge. By WILLIAM POTTS. *Macmillan's Miniature Series.* 18mo, 75 cents.

"But the attraction of Mr. Potts' book is not merely in its record of the natural year. He has been building a house, and we have the humors and the satisfactions, and hopes deferred, that usually attend that business. He has been digging a well, and the truth which he has found at the bottom of that he has duly set forth. . . . Then, too, his village is Farmington, Conn., and there Miss Porter has her famous schools, and her young ladies flit across his page and lend their brightness to the scene. And, moreover, he sometimes comes back to the city, and he writes pleasantly of his New York club, the Century. Last, but not least, there are lucubrations on a great many personal and social topics, in which the touch is light and graceful and the philosophy is sound and sweet." — *Brooklyn Standard-Union.*

WEED. — **Life Histories of American Insects.** By Professor CLARENCE M. WEED, New Hampshire College of Agriculture and Mechanical Arts. Fully Illustrated. Cloth. $1.50.

THE MACMILLAN COMPANY,
66 FIFTH AVENUE, NEW YORK.

Printed by Amazon Italia Logistica S.r.l.
Torrazza Piemonte (TO), Italy

44150458R00198